NEW
THINKING

I would like to thank God, my family, and my friends for their support, and for tolerating my occasional complaining, while writing this book. A special thanks to my sister, Piriye, for her role in helping this book take shape. Last but not least, I want to show gratitude to all of my YouTube subscribers for making this opportunity possible in the first place.

ColdFusion Presents:

NEW THINKING

From Einstein to Artificial Intelligence, the Science
and Technology That Transformed Our World

Dagogo Altraide

Mango Publishing
Coral Gables

Cover Design: Dagogo Altraide
Layout Design: Jermaine Lau

Images from shutterstock:

shutterstock.com/ID778268278, shutterstock.com/ID615282242, shutterstock.com/ID644426005, shutterstock.com/ID1092097760, shutterstock.com/ID339962852, shutterstock.com/ID252140368, shutterstock.com/ID203537950, shutterstock.com/ID31697947, shutterstock.com/ID786721612, shutterstock.com/ID217940149, shutterstock.com/ID249574177, shutterstock.com/ID244389040, shutterstock.com/ID1040827711, shutterstock.com/ID33568546, shutterstock.com/ID1092563384, shutterstock.com/ID438296326, shutterstock.com/ID610775501, shutterstock.com/ID1080951473, shutterstock.com/ID1155148708

All other images created by Mango Publishing Group and/or provided by Dagogo Altraide

For permission requests, please contact the publisher at:

Mango Publishing Group
2850 Douglas Road, 2nd Floor
Coral Gables, FL 33134 USA
info@mango.bz

For special orders, quantity sales, course adoptions and corporate sales, please email the publisher at sales@mango.bz. For trade and wholesale sales, please contact Ingram Publisher Services at customer.service@ingramcontent.com or +1.800.509.4887.

Cold Fusion Presents: New Thinking: From Einstein to Artificial Intelligence, the Science and Technology That Transformed Our World

Library of Congress Cataloging
ISBN: (print) 978-1-64250-591-7 (ebook) 978-1-63353-751-4

Library of Congress Control Number: 2018952299

BISAC category code: "TEC000000—TECHNOLOGY & ENGINEERING / General"

Printed in the United States of America

Table of Contents

Introduction

*"The spread of civilization may be likened to a fire;
first, a feeble spark, next a flickering flame, then a
mighty blaze, ever increasing in speed and power."*

—Nikola Tesla

The history of mankind is built on new thinking. We use the previous generation of tools as a foundation to build new tools, which in turn build even more powerful tools—a feedback loop that keeps on accelerating. Some of these innovations change the world forever: fire, steam power, the transistor. Some, not so much: the Power Glove, the Clapper, Smell-O-Vision. What hidden stories lie behind the technology we all use today? *New Thinking* is the story of human innovation. Through war and peace, it is humanity at its most inventive, and sometimes most destructive. In this book, we will take a walk through the history of technology, the history of us: from the Industrial Revolution to Artificial Intelligence.

Before we start, however, we need to talk about the new thinker of all new thinkers—an inventor who is so important to the history of technology that he's nicknamed "The Man Who Invented the Twentieth Century." I am speaking, of course, about Nikola Tesla.

You won't find Tesla's name in the following chapters of this book. This isn't because he isn't important. It's because he is too important. If I were to include Tesla, his name would be on every second page.

Let's run down a few of the inventions he was either instrumental in realizing, or invented himself: alternating current, the induction motor, the Tesla coil, wireless lighting, the steam-powered oscillating generator, the radio, hydroelectric power, x-ray imaging, and the remote control. This doesn't even begin to list the things he envisioned but couldn't get around to or realize because his thinking was too advanced for the materials at hand. There is a reason Elon Musk named his car company after Tesla.

The legend of Nikola Tesla grows by the year, and the crazy thing is, the legend probably doesn't even capture half of the amazing truth. This is a man who once built a small earthquake machine in New York, and then dared Mark Twain to stand on it. The machine only caused a small rumble, but it was enough to loosen the bowels of the famous author. Tesla's legend includes wireless power, weather control, and a death ray that he reportedly carried with him in an unmarked bag—a bag that mysteriously went missing after his death.

Many of the wilder legends about Tesla are unsubstantiated, though there are more than enough verified stories to fill an encyclopedia. Tesla was so ahead of his time that, when he first displayed his radio remote-control boat at an electricity exhibition in Madison Square Garden, the technology was so far beyond anything onlookers had seen that some literally thought Tesla was either a magician or telepathic, while others chalked the display up to a tiny trained monkey that had been hidden in the remote-control boat.

Alternating current, along with the induction motor, is the reason we can plug things into the walls in our homes. It was such a huge step that it wasn't just shown off at the famous 1893 Chicago World's Fair, it was used to light it.

Without Tesla, we wouldn't have the electricity in our homes, the motors in our cars, or the ability to change the channel when *American Idol* starts. We wouldn't be able to see broken bones on an x-ray image or listen to the weekly Top 40 on the radio.

Tesla is the poster boy for Arthur C. Clarke's famous quote: "Any sufficiently advanced technology is indistinguishable from magic."

But rather than spending all of our time with this amazing Serbian-born inventor, let's start at the beginning with chapter 1, and the Industrial Revolution.

PART 1

Origins

CHAPTER 1

The Industrial Revolution

If you've ever eaten food you didn't grow, put on clothes you didn't make, driven a car, used electricity, watched TV, used a phone or computer, slept on a bed, used a toilet, consumed water from a tap, or been inside a building, then congratulations, you've lived with the consequences of the Industrial Revolution. This event was the single biggest change for mankind in history.

Before the Industrial Revolution, people lived on the land that provided them with food and the means for clothing and shelter. Average life expectancy was around forty years (including infant mortality), and any form of structured education was extremely rare. All the while, disease and malnutrition were rife. Until the revolution, humans never used tools or objects that weren't produced within their immediate community or traded. The fastest any human could travel was the speed of a horse. Over 80 percent of the population lived on farms. With no mass production or the ability to transport large quantities of goods a long distance, people had to be close to their source of food. It was the only means of living. Today, the number of people on farms in the United States is down to less than 1 percent.

So where did the dawn of our modern era take place?

The Steam Engine That Powered a Revolution

It all began in England, around 1712. At the time, a primitive tin and coal mining industry existed, but there was a major problem. The mines would get flooded whenever it rained, and in England, rain was a pretty common occurrence. Every time a mine flooded, production stopped. This meant that production was subject to weather conditions. To deal with the flooding, scores of men carrying endless buckets of water would be commissioned to bail out the mines. As you can imagine, this was very inefficient and costly. There had to be a better solution.

Enter Thomas Newcomen from Dartmouth, Devon. He was the inventor of the first practical steam engine.

What's important here is the revolutionary application of one of water's most fundamental properties: that heated water turns to gas (steam), and that this steam expands, pushing objects in its wake, causing motion. This was the first practical device that used steam to produce motion.

Newcomen Atmospheric Engine

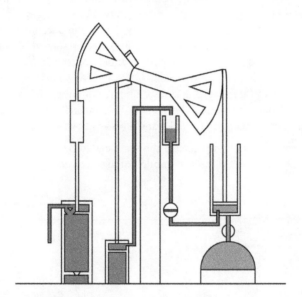

This steam pump (now named the Newcomen engine) was put to use in the mines. This in turn increased the production of coal and tin. Through the power of steam, human effort was no longer needed to bail out the flood water. As great as this was, there were some problems with Newcomen's engine. It was slow and used a lot of coal, making it expensive to run.

James Watt

With a basis to start from, there was now room for someone to come through and improve the technology of the Newcomen steam engine. That someone was Scotsman James Watt, the man who truly got the revolution going.

Born on January 19, 1736, James Watt was the son of a shipyard owner. While in school, he was taught Latin and Greek, but was thought to be "slow" by his teachers. As it turned out, he just wasn't interested in language. When it came to engineering and mathematics, however, James excelled.

At age nineteen, he went to Glasgow to study the trade of making precisely calibrated instruments such as scales and parts for telescopes. Watt eventually made instruments for the University of Glasgow. During his time at the university, Watt was given a model of a Newcomen engine to repair. Very quickly, he became interested in steam engines and noticed how inefficient the standard Newcomen engine of the day was. He decided he could improve it. While taking a walk on a Sunday afternoon in 1763, an idea struck him. Instead of heating and cooling the same cylinder, why not have a separate chamber that condenses the steam? This meant that the machine could work in both an upstroke and a downstroke motion.

This idea would end up cutting the fuel needed by 75 percent. After experimenting with a small model of his new design, Watt was

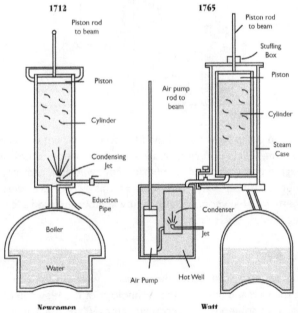

convinced it would work. Soon, a partnership with an industrialist by

the name of Matthew Boulton was formed. This partnership would
alter the world for good.

It has been said that Boulton was a little like Steve Jobs, an enthusiastic, business-minded individual, while Watt was like a gloomy version of Steve Wozniak—the man behind most of the technical work. This isn't far from the truth, although there was much more crossover in the roles of Watt and Boulton than with Apple's founders.

Throughout the mid-1770s, James Watt and Matthew Boulton would use their own company (Boulton & Watt) to distribute the new steam engines throughout England. The impact was immediate within the mining industry, and it also reached the liquor industry with grinding malt.

To explain the benefits of the machine, Watt had to come up with a way of relaying its power. He figured that a horse could pull around 82 kg (180 pounds), so, in his description to customers, Watt would say, for example: my machine has the power of twenty horses—twenty horsepower. This is in fact where the unit of power came from.

Subsequent improvements in the steam engine soon opened the door to powered factories and a revolution in the textile industry. For the first time, the mass production of goods was possible. These conditions allowed for new employment opportunities in city locations. As a result, job seekers left their farms and headed to the city in search of a new life.

Steam Revolutionizes Transport

Steam power had now revolutionized production, but Watt realized that, by expanding the gas at even higher pressure, this invention could be used in transportation. The locomotive application for the steam engine would push humanity to another level. The first patent of this kind was obtained in 1784 by Watt, though it is often said to be the brainchild of Boulton & Watt employee William Murdoch. These patents barred anyone from creating higher-pressure versions of the Watt steam engine until 1800. When the patent expired, the floodgates were opened, and innovation flowed.

One of the first improvements was made by Cornish engineer, Richard Trevithick, who enabled the use of high-pressure steam. Yielding more power, this development opened the door to feasible locomotive steam engines. Improved designs and power-to-size ratios meant that engines became so compact they could be used, not just in factories, but also in mobile machines.

The year 1804 was monumental in history. That year would see the world population reach one billion, the isolation of morphine, and Napoleon come to power; but most of all, it was the first time goods were transported over land without the power of man or animal. This feat came in the form of a steam locomotive with a speed of 8 km an hour (5 miles per hour) carrying a load of more than twenty-five tons. Not bad, considering cars were still almost a century away.

Steam-powered trains and railroads became a major British export and began to have a small impact on the rest of Europe. Arguably, however, the biggest effect was seen in the United States. In the early 1800s, many models of locomotives were imported from Great Britain, but by 1830, the United States was building its own trains. American companies began forming and a new industry emerged.

Steam Transforms America

At first, the tracks were no more than a few miles in length, so long-distance rail travel became the holy grail. Previously, Americans had tried camel caravans and horse-drawn stagecoaches to deliver mail or travel over long distances, but these attempts had met with limited success.

A trip from St. Louis to San Francisco, via either the camel-caravan or stagecoach method, would travel 2,800 miles (4,500 km) of dirt trails and last around three weeks. American writer Mark Twain went on one of these stagecoach trips. He was unpleasantly surprised by the experience.

The meals consisted of beans, stale bacon, and crusty bread. He described the comfort level as "bone-jarring," "teeth-rattling," and "muscle-straining."

By 1863, it was time for a change. A young civil engineer by the name of Theodore Judah had a vision to build the ultimate railway, a railway so large it would connect America from the west to the east. Around that time, members of the United States Congress were thinking about such a railway but couldn't determine the precise route on which it should be laid. Judah figured out the perfect route and stepped in as the one to build the tracks. He contracted a company called Union Pacific to build from the East Coast and another company, Central

Pacific, to build from the West. On May 10, 1869, after six years of hard work—including laying steel tracks in the Nevada desert and the devastating sacrifice of much human life—the two companies met in Utah, and the first transcontinental railroad was built.

Thanks to this sacrifice and hard work, California was now connected to the rest of the nation.

With steam-powered locomotion, people and large amounts of cargo could travel long distances across land, with relative ease, for the first time. The possibilities were endless.

Goods and services could be transported to support new towns that weren't by ports. It became less common for people to be born and to die in the same place—the common man was now mobile. The California connection allowed perishable food to quickly be transported across the country in refrigerated railcars, ushering in a new era of prosperity.

We essentially still use steam engines as a way of generating power today. Coal, nuclear, and some natural-gas power plants all boil water to produce steam. This steam then drives a turbine that generates electricity. It's amazing that the consequences of Watt's idea during a Sunday stroll still impact us today.

While the steam engine was just starting to move from the factory onto the railroad, there was another technology whose time had not yet arrived; however, it would soon be just as, if not more, revolutionary.

This innovation was none other than electricity.

CHAPTER 2

Building a Foundation

THE MYSTERY OF ELECTRICITY

From ancient times, electricity was known only as an abstract concept. Imagine you were living five thousand years ago and received a shock from touching an electric fish. There would have been nothing like it in your life experience, except perhaps for lightning. The fish would seem mysterious, perhaps even magical. This enchanting puzzle existed for centuries and millennia.

The term "electricity" came from William Gilbert in his 1600 publication De Magnete, considered to be the first real scientific work published in England. The word "electricity" was derived from the Greek word for amber: ἤλεκτρον (elektron).

Alessandro Volta Invents the Battery

The end of the century would see man create his own form of electricity. In 1799, Italian inventor Alessandro Volta developed the voltaic pile, a stack of zinc and copper discs separated by salt-water-soaked cardboard. The device was the first to provide an electric current. It was essentially a battery.

In 1800, an excited Volta reported his findings in a letter to the Royal Society in London. It was a bombshell at the time, because the governing scientific consensus was that electricity could only be generated by living beings. It sparked great enthusiasm. It soon opened the door to the isolation of new chemical elements, such as potassium, calcium, and magnesium, through electrolysis.

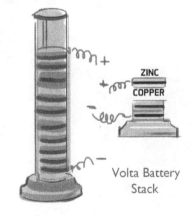

ZINC
COPPER

Volta Battery
Stack

After his discovery, Volta was held in great esteem by many, and even had the unit of electric potential (the volt) named after him. Despite this fame, Volta was a dedicated family man and often shielded himself from the public eye for this reason.

The voltaic pile built a foundation for electricity. However, the time for widespread electricity had not yet come.

A False Start for Electrical Lighting

For the longest time, oil lamps and candlelight were the only ways people could see what they were doing after the sun had set. If you were out running an errand and ran out of lamp oil on the way back home, too bad. This was set to change around 1805, when an Englishman by the name of Humphry Davy demonstrated light by electricity. His invention was a type of lamp that gained its illumination from running electricity through small carbon rods. The rods were separated by a 4-inch (100-mm) gap, and when the electricity passed across the gap, it would form an "arc" of white-hot light. This form of lighting was to be known as "arc lighting," usually facilitated by high-voltage batteries made from thousands of Volta cells.

Arc lamps were extremely bright for their day. Some estimates put them at an equivalent brightness of 4,000 candles (about ten times less than a modern car's focused headlight).

With an arc light's brightness came a lot of heat, and with that heat came the risk of fire. Because of their brightness and potential to be a fire hazard, they were deemed unsuitable for the home.

Arc lights were used in commercial settings, but there was no steady supply of electricity to power these lamps for long periods of time. It was a dead end for the widespread use of electric lighting. Electric lighting does make a comeback later in the chapter, as we progress through the century.

Meanwhile, other applications of electricity were about to make some big strides.

Michael Faraday

Michael Faraday is a name that might be familiar to many. He was one of the most influential scientists in history, revolutionizing our understanding of electricity, chemistry, and electromagnetism. His achievements include the invention of the Bunsen burner and the DC electric motor, the discovery of benzene, and the development of the Faraday cage. Albert Einstein even kept a picture of Faraday on the wall in his office.

Surprisingly, Faraday received little formal education after his birth in England in 1791. He had to educate himself. At around age fifteen, while working as a bookbinder, Faraday came across two titles that would change his life and the course of history. One was *The Improvement of the Mind* by Isaac Watts, and the other was *Conversations on Chemistry* by Jane Marcet. The first gave Faraday the tools of understanding and a disciplined mind, while the latter inspired him to move in the direction of electrical experimentation. After attending the lectures of English chemist Humphry Davy at the English Royal Society, Faraday ended up going on a European tour with him to serve as an assistant.

Unfortunately, Faraday was looked down on at the time and was not considered a "gentleman." He was made to sit on top of the stagecoach, rather than within it, and eat with the servants when meal breaks came. This upset him greatly, but the opportunities to meet great intellectuals and his passion for learning more about the world of science kept him going.

Faraday's genius was in the way he expressed his ideas. It was done in clear and simple language. That was really the only way it could be done, as his knowledge of math was limited. Despite Faraday's limited math, he managed to liquefy gases, produce new kinds of glass, improve steel alloys, and discover nanoparticles and benzene, which is currently used in the production of plastics, dyes, drugs, lubricants, rubber, and more. However, none of these were his most brilliant discovery. That honor goes to the 1821 use of a voltaic pile (Volta's battery) to produce motion—in other words, a motor.

Davy and two of the best scientists of the day (including James Maxwell, who discovered electromagnetism) had previously tried, but failed, to build an electric motor. In 1821, it was discovered that electricity could actually produce a force when in the presence of magnets.

Faraday, who was relatively uneducated compared to these men, successfully produced a continuous-circular-motion motor after simply discussing this problem with them. It only generated a very weak force, but it worked.

Think of the parallel here between electricity driving motion and steam driving motion. Again, we see how the ideas of those before are built on by those after, producing new technologies.

Today, although they look vastly different from Faraday's motor, the very same principles discovered by Faraday are used in our computer hard drives, electric bikes, and phone vibration motors. Notably, Nikola Tesla improved on Faraday's idea to create the AC motor, the kind used in everything from washing machines to electric cars.

But Faraday wasn't done yet. In 1831, he created the electric generator, otherwise known as a dynamo. This device did essentially the opposite of the electric motor: It turned motion into electrical energy.

By the 1830s, the mystery of electricity was starting to be understood, but it wasn't revolutionary just yet. This was because the dynamo produced a fairly low amount of voltage (only a few volts) and was inefficient. Although it proved that electricity could be generated by using magnetism, there were severe limitations on what it could power. Regardless, the new thinkers still sought uses for this latest invention.

ELECTRIC COMMUNICATION BEGINS

Unlike the previous inventors we've covered, Charles Wheatstone (born in England in 1802) came from a musical background. His father sold instruments and was a flute teacher. As a teenager, he was interested in a wide array of books, and often spent time at an old bookstall in his local mall. One day, he came across a book that covered the discoveries of Volta, who had invented the battery just over a decade earlier. He didn't have any money, but soon saved up to buy the book.

Charles Wheatstone and his brother decided to make a battery of their own. They soon realized that they weren't going to have enough money to purchase copper plates for its construction. Charles was dedicated enough to use his remaining copper pennies to complete the battery.

In 1821, Wheatstone attracted attention in his local mall by creating a sound box that relayed the sounds of live instruments, played over a wire—rather amusing for the day. He would also coin the term "microphone" for an invention that amplified into each ear mechanically.

The Electric Telegraph

By 1835, Charles saw the potential of using electricity to transmit information through his work as a professor at King's College London. This realization caused him to abandon his mechanical microphone methods and explore the concept of the electric telegraph. He could see that the technology, in its perfected state, held incredible potential for the entire world. The idea was to build on established, primitive methods of electrical telegraph communication—most notably the work of Russian inventor Baron Schilling, who managed to communicate between two rooms. Electric telegraph messages could work on very little voltage, making them an ideal technology for the power generation available at the time.

Meanwhile, William Cooke, an officer in the Indian army, was on leave and decided to attend some anatomy lectures at the University of Heidelberg. While there, he saw a demonstration of a primitive electric telegraph and immediately realized its importance. Cooke dropped his anatomical studies and began to look into the telegraph. For consultation on scientific knowledge, he contacted Michael Faraday and Peter Roget, an academic in Edinburgh. Roget suggested that Cooke pay a visit to Wheatstone.

Cooke proposed a partnership with Wheatstone, but the two had very different motives for the project. Wheatstone was purely interested in the scientific significance of the technology, while Cooke was interested in making a fortune. Despite this, after some hesitation from Wheatstone, the partnership was agreed upon.

In 1837, the pair set up an experimental machine between the Euston and Camden Town train stations in London, a distance of 2.4 km. The receiver of the telegraph message was to read it via five electromagnetic needles that pointed toward an array of letters in a diamond configuration. To send and receive messages, the system needed at least five wires to transfer the information.

On July 25, 1837, the first message was sent. Wheatstone recalls his feelings at the time: "Never did I feel such a tumultuous sensation before as when, all alone in the still room, I heard the needles click, and as I spelled the words, I felt all the magnitude of the invention pronounced to be practicable beyond cavil or dispute."

The railway directors, however, were not impressed. They saw the telegraph as a silly contraption, nothing more than a foolish plaything. They wanted it promptly removed from the station. All seemed lost, but two years later, the invention was picked up again by England's Great Western Railway for a 21-kilometer distance—marking the first practical use of the telegraph.

Any Publicity is Good Publicity

The telegraph was still relatively unknown to the public, until the most unusual of promotions for any sort of technology—an 1845 criminal case involving a murderer on the loose. The suspect's name was John Tawell, and he had boarded a train to flee toward London. Once his destination was known, a telegraph message was sent to his intended arrival station. When Tawell stepped out of the carriage, the police were already there to arrest him. This event was astonishing in its day. A form of communication so fast, it was able to intercept a criminal on a moving train? There was great publicity around the event, and the usefulness of the telegraph was finally established.

Over in the United States, a series of tragic events would push another inventor to think about improving communication even further.

Samuel Morse

Imagine this scenario: It's 1825, and you're in Washington, DC, creating a painting of a high-profile public figure. In New Haven, Connecticut, 300 miles (480 km) away, is your sick wife. One day, as you're working on the painting, you receive a message from a horse-riding messenger saying that your wife's health is improving. You paint on. The very next day, you receive another message by horse messenger. This time, it's a detailed report of your wife's death. Overcome with sorrow, you abandon everything and leave for New Haven. By the time you arrive, your spouse has already been buried.

This is exactly what happened to Samuel Morse.

He was plunged into great despair by these tragic events, but it would serve as the catalyst for him to begin his search for rapid long-distance communication. The result was the single-wire telegraph and

Morse code, the latter of which became the standard for telegraph communication and is still used today.

Morse code is an ingenious and simple language that uses dots and dashes to stand for the letters of the alphabet. A skilled operator could send eighty characters a minute.

Morse sent his first message from Washington, DC, to Baltimore, Maryland, in 1844. It said: "What hath God wrought?" It was taken from Numbers 23:23 in the Bible. Morse code became the standard for the telegraph, as it was simpler to send a message in dots and dashes than to use at least five wires corresponding to letters, like the Wheatstone machine.

Perhaps, if tragedy had never struck, Morse would never have been known for more than his great works of art.

Going the Distance

Electricity had allowed for rapid communication between cities, but for nations separated by seas, communication was still only possible by ship. This was very slow. For example, a message sent between the United States and England would take over ten days to arrive.

In the 1850s, as the buzz around the telegraph was growing, there was an increasing consensus to apply this technology across the Atlantic Ocean (around 2,000 miles, 3,200 km). In 1854, the Atlantic Telegraph Company began construction of the first transatlantic telegraph cable. British and American ships were used to lay down an insulated cable on the ocean floor. After four failed attempts, the project was completed in 1858, on the fifth attempt. It was the first project of its kind to succeed. Despite all this effort, the cable functioned for only three weeks.

The first official telegram to pass between two continents was transmitted on August 16, 1858: a letter of congratulations on completion of the project, from Queen Victoria to the United States president, James Buchanan. Soon after, signal quality deteriorated, and transmission became sluggish. In an attempt to speed up communication, the White House pushed extra voltage through the cable, destroying it completely. When the line was functional, it had brought British-US communication time down from ten days to about seventeen hours.

In the United States, telegraph lines were already connecting cities throughout the east of the country. Work began in 1851 on the task of building a telegraph line across the country. On October 24, 1861, the first telegram was sent from San Francisco to Washington, DC. It would also be the day the Pony Express went out of business. The Pony Express was previously the fastest method of getting mail (information) from the east to the west, as the railroad was still eight years away. The electric telegraph rendered the Pony Express obsolete, just as steam locomotives eventually rendered horse-coach transport obsolete.

Written communication was now possible over long distances in short periods of time, with vast economic and social impact, but what about speech?

ALEXANDER GRAHAM BELL: THE TELEPHONE

Scotsman Alexander Graham Bell was born in 1847, right around the time the telegraph was building up momentum in England and the United States. Bell may be remembered for his invention of the telephone, but—as we'll see later in the book—his legacy would directly build the foundation for computing through the invention of transistors.

Bell was a curious child and had a natural knack for invention. At age twelve, he built a device to husk wheat at his best friend's family's flour mill. Although he had a mind for practical invention, he did poorly in school and left at age fifteen; however, he did pursue higher education at a later date.

Bell was an introvert who was especially close to his mother. In his boyhood, she encouraged him to develop his natural talent for art and music. When she began to lose her hearing, Bell was hit hard. These events inspired him to study acoustics. He would eventually run his own school for the deaf in Boston.

Alex and His Tricks

Bell would often perform creative and playful experiments with speech to entertain others. Once, he used the family dog for one of those experiments. He trained the dog to maintain an extended growl. While the dog made the sound, he would manipulate the dog's lips and vocal cords to produce a phonetic "ow ah oo ga ma ma." With a little convincing, visitors believed his dog could articulate "How are you, Grandmama?" Witnesses were adamant that they were seeing a talking dog!

Bell also gave lectures on "visible language" to teachers of the deaf, but soon invented his own methods of lip reading and teaching speech. By 1871, he was a leader in the field of hearing-disability education. It was his view that the deaf should be liberated from sign language and should be able to speak. At the time, deaf individuals were usually excluded from society.

Bell began working on a harmonic telegraph around this time. He used his musical background and knowledge of sound to realize that multiple signals (Morse code dots and dashes) could be sent through the same wire if they were at different pitches.

Here's how it worked: a telegrapher could tap out the message in Morse code. The signal of the code was then transmitted as short audio beeps of a specific frequency. On the same line (wire), another telegrapher could tap out a Morse code message at a different frequency. At the receiving end, a device would only respond (vibrate) to one frequency, not to others on the same wire. By using many receivers set to different frequencies, multiple messages could be sent across the same wire without getting mixed up. Bell called this a multi-signal telegraph.

After securing financial backing from two close friends, Bell further developed the idea with a young assistant, Thomas Watson. One day, while Bell and Watson were working on the project, Watson accidentally plucked one of the transmitting devices. Instead of hearing a beeping tone like that from a Morse code message, Bell heard an audible twang through the receiver in another room. At once, Bell realized that, if a twang could be transmitted through this device, there was potential to transmit *any* sound—even voice—over wire electronically. This was a watershed moment. It marked the birth of the telephone and the death of the multi-signal telegraph. The date was March 10, 1876, and on this day the first words ever spoken by telephone were vocalized: "Mr. Watson, come here. I want to see you."

On August 4, 1876, after creating a working model of the telephone, Bell gave a demonstration to his friends and family. They were amazed when they heard the voices of people reading and singing at another location 6 km (about 4 miles) away.

The telephone would get its international break later that year, when Bell set up a demonstration of the prototype machine at the Centennial Exposition in Philadelphia. Viewers of the demonstration were prominent people of the day. One such person was Queen Victoria. She called the invention "most extraordinary."

Not everyone was as impressed, however. A short time after the demonstration, William Orton, president of Western Union, had the opportunity to buy Bell's invention. He infamously said: "What use could this company [Western Union] make of an electrical toy?" Orton, of course, was badly mistaken.

With the wind in Alex's sails (aside from Orton), he set up the Bell Telephone company one year later, in 1877. The Bell Telephone Company would later be known as the American Telephone and Telegraph Company, or AT&T, still one of the largest service providers today.

Less than a decade after Bell's public demonstration, there were more than 150,000 users of the telephone in the United States.

The telephone fulfilled one of the greatest human needs, allowing people separated by vast distance to hear and speak to each other. Indeed, it was one of the most successful inventions of all time. In fact, our smartphones today are the great-grandchildren of Bell's invention.

What most people don't realize is that Alexander Graham Bell often reminisced that his greatest contribution to the world was not the invention of the telephone, but his work of oral education for the deaf.

By the late 1870s, electricity had made it possible for humans to communicate using their voices (rather than Morse code signals) across large distances. But was there a way to preserve these sounds? At this point, there was still no way to record sounds and play them back later. Thomas Edison would crack the code.

Thomas Edison: Recorded Sound

Edison stumbled upon the solution in July of 1877, while he was working on a way to record and play back telegraph messages. He first did this by capturing Morse code bursts as vibration-induced indentations on a spool of paper. After some thought, he theorized that the same could be done with any sound. All he needed was a vibrating object that could mark its motion onto an impressionable surface. The first method used a cylinder wrapped in foil.

Edison's unveiling of his invention to the press was truly the work of a clever mind. A writer from the *Scientific American* recalls the event: "In December 1877, a young man came into the office of the *Scientific American*, and placed before the editors a small, simple machine about which very few preliminary remarks were offered. The visitor without any ceremony whatever turned the crank, and to the astonishment of all present the machine said: 'Good morning. How do you do? How do you like the phonograph?' The machine thus spoke for itself, and made known the fact that it was the phonograph..."

The patent for the phonograph, US Patent No. 2,000,521, was awarded on February 19, 1878. The technology would be improved by Alexander Graham Bell, who, after listening with a keen ear, thought to replace the tinfoil cylinder with a wax cylinder. The resulting sound was much better, leading to the commercialization of the first recorded music in the 1890s.

ELECTRIC LIGHT'S TURN TO SHINE

In the 1870s, work was still being done on Michel Faraday's electric generator (the dynamo we saw earlier, which turned motion into electrical energy). By the end of the decade, other nations became interested in the technology, and tests were being carried out to link this electric generator to the arc-lighting concept.

The first practical large-scale generators were produced in France, resulting in the first electric street lighting on the planet in 1875. By 1878, Paris had arc-lighting systems installed around the city. A writer for a London technical journal, *The Electrician*, visited France and was impressed by the feat, but slightly bitter that this technology had not come to Britain. The journal noted: "The application of the electric light is in Paris daily extending, yet in London there is not one such light to be seen."

In response, further advances in electricity generation were made by the British, and, by the start of the 1890s, seven hundred arc lamps were in use in the country. Markets, railroads, and the curbs in front of large buildings were ideal locations to be fitted with arc lighting.

Regardless of this use in public settings, there was no such electrical lighting for the private home.

Thomas Edison, Light, and Controversy

The idea of the filament lamp was almost as old as the arc lamp. Many people had tried to produce light by heating a fine wire electrically so that it glowed (producing incandescent light), but it seemed impossible thus far. There were several reasons for this: First a material had to be found that could withstand being repeatedly heated and cooled; second, this material had to be sealed inside a glass vessel such that the glass did not crack when the wire was hot; and third, the air had to be pumped out, so the filament did not oxidize (rust).

Today, many associate Thomas Edison with the invention of the first incandescent lightbulb, but, like many of Edison's "inventions," his work involved less new thinking and more improving on existing ideas. There are in fact dozens of examples of incandescence being used for light, dating back to the 1700s, but it wasn't until the late 1870s that the light bulb race really heated up. During that time, the key competitors hailed from either side of the Atlantic: there were Thomas Edison and William Sawyer competing in New York, and Joseph Swan competing in London.

Swan led the pack early. He experimented with a number of materials, including carbon, platinum, and cotton thread. He successfully obtained a British patent for his work, along with a list of firsts, including: first to light (produce lighting for) houses, first to light a theatre, and first to light a public building.

Meanwhile, in New York, Edison filed for his own patent on "Improvement in Electric Lights." Edison had also begun with carbon and platinum filaments before his team discovered that carbonized bamboo could last for over 1,200 hours. The bulbs created by Edison's team were no doubt groundbreaking—though it was suspicious that the work seemed to be largely based on designs by fellow American (and competitor) William Sawyer.

A usable light bulb had been invented, but by whom? There were now a lot of conflicting patents. Edison lost a suit in England, forcing him to cede his claim to Swan. Meanwhile, back in the US, William Sawyer's partner Albon Man was challenging Edison's patents on his home soil. Edison initially lost the case, but, after six years of fighting in the courts, he finally ended up on top. A few years later, he also managed to buy Swan's UK company out from under him. From this time on, the legend of the Edison light bulb solidified in history.

By 1900, a few great minds had given us the telephone, recorded sound, electric light, and trains. It really was a new world.

CHAPTER 3

A New World
1900-1909

The year 1900 seems like an eternity ago. The world then was a completely different place. In 1900, the US population was 75 million, L. Frank Baum's book *The Wonderful Wizard of Oz* was newly released, and work on the New York subway system was just beginning.

In its introductory decade, the twentieth century saw many firsts. The 1900 Paris Olympic Games was the first in which women could compete. 1901 saw the first ever Nobel Prize ceremony. 1902 marked the year of the first theater devoted solely to films—the Electric Theatre in Los Angeles. Four years later, the first feature-length film, *The Story of the Kelly Gang*, would be produced in Melbourne, Australia. A few other inventions of the decade were the washing machine, the vacuum cleaner, and the first entertainment radio broadcast.

Also, in 1901, Rudolph Diesel unveiled his diesel-design engine (running on peanut oil) at the World's Fair in Paris. The World's Fair had 50 million attendees and showcased other emerging technologies, such as "talking films," escalators, and the first magnetic audio recorder. At the time, Paris was a hub for technological innovation.

The borders of nations were also shifting as displays of power were made. The new imperialism was in full effect: The powers of Europe, the United States, and Japan were all trying to literally take over the world by owning ever-increasing areas of territory in Asia, Africa, and South America.

Meanwhile, great minds were at work: In 1905, Einstein's theory of special relativity was released to the world. It postulated that the laws of physics are the same everywhere for a stationary observer, and that the speed of light is constant. The theory hypothesized odd things—like light bending due to gravity, and black holes. Perhaps one of the strangest relativistic effects is what happens to an object as it approaches the speed of light. The closer to the speed of light the object travels, the heavier it gets. An object traveling just under the speed of light will be infinitely heavy. Einstein's theory also had some strange side effects, such as time itself being different depending on where you are and how fast you're going. For example, time slows down as one approaches the speed of light, an effect called time dilation.

This might sound far from everyday practical use, but special relativity is used every day. For example, if the clocks on our GPS satellites traveling around the globe at 10,000 km an hour (6,000 miles per hour) didn't adjust for time dilation, their measurements would be off by 8 km (5 miles) after just one day! Imagine trying to navigate using Google Maps with that kind of accuracy.

To have a mind that could foresee this effect in 1905 is incredible, but apart from Einstein, there were other great minds in the 1900s, working on ideas that impact our world today.

THE DAWN OF PERSONAL AUTOMOBILE TRANSPORT

The decade kicked off with a bang, or more specifically a sequence of very fast bangs. In the context of technology, the first few decades of the 1900s saw the widespread introduction of the internal combustion engine. In a way, this technology took the baton from the steam engine of the 1800s. You may be surprised to know that, during the 1800s, there were in fact steam-powered cars. These cars took up to 45 minutes to start up in wintry weather and had to be topped up with substantial amounts of water, both factors which limited their range. There had to be better solutions.

And there were. The first hydrogen car appeared in 1808, while 1838 saw the first primitive electric car, and 1870 ushered in the first gasoline car. In 1885, Karl Benz produced the first gasoline-powered car that wasn't just a one-off (Benz's company would eventually become Mercedes-Benz).

A Lot of the First Cars Were...Electric?

Early gasoline-powered cars weren't perfect, however. To get them started, you needed a lot of muscle to wrestle with the hand crank; even changing gears was complicated. The throttle was actually on the steering wheel. For these reasons, electric cars were seen as the best way forward for personal transport. They were whisper-quiet, easy to drive, and required relatively little effort to run.

Big names such as Ferdinand Porsche (founder of the Porsche motor company) and Thomas Edison went all-in on electric. Porsche's very first car, in 1898, was an electric model. One year later, Edison set out on a mission to make longer-lasting batteries for cars. He believed electric transport was clearly the future. Despite many attempts, he ultimately abandoned this vision a decade later.

Electric cars were perfectly fit for short trips around the city, as country roads weren't suited for cars at the time. By 1900, only 22 percent of cars were powered by gasoline, while 40 percent were electric, and the remaining 38 percent ran on steam.

At the turn of the century, the car (electric or otherwise) was seen as an expensive novelty. Only royalty and the very rich could experience the benefits of personal automatic transport. To have a world where the common man could travel as he pleased required a disruption in the way cars were produced. Cars needed to be affordable. It was a man named Henry Ford who would bring this into reality.

Henry Ford: The Father of the Affordable Automobile

In 1891, the Edison Illuminating Company took on a bright twenty-eight-year-old engineer. He was a natural, and quickly rose up the ranks. In just two years' time, he was Chief Engineer. His name was Henry Ford.

Henry Ford was born on July 30, 1863, in Michigan. When he was thirteen, his mother passed away and his father decreed that Henry was to take over the family farm. Although he had great love for his mother, he despised the idea of a life of farm work. Ford showed interest in mechanical tasks as a child, and, when he was around fifteen years old, he would repair watches for friends and neighbors.

At sixteen, Ford left the farm and worked in Detroit as an apprentice machine operator before being picked up by the Edison Company in 1891. In his spare time at the company, Edison would test out ideas for a horseless carriage. In 1896, he constructed his first car, cleverly named the Ford "Quadracycle." He would present the plans of his new vehicle at meetings with Edison company executives. Interestingly, Edison himself was eager to see another improved model from Ford. Three years later, Ford left the Edison company to work full-time on building a car.

After two failed attempts, the Ford Motor Company was established in 1903. The very first Ford was the Model A. It featured an 8-horsepower motor (horsepower—remember James Watt's term from chapter 1?) and sold for about $800–900, depending on whether you went for the two-seat or four-seat version. 1,750 cars were made from 1903 through 1904, all in one color—red (perhaps because it made the car a little faster). The company had spent almost its entire $28,000 in initial investment funds and had only $223.65 left in its bank account when the first Model A was sold in 1903. It was make-or-break for the company at that moment, but fortunately the American people loved the car, and it went on to be a success.

Who Owned the Very First Ford?

On July 23, 1903, Chicago dentist Ernest Pfennig became the first owner of a Model A. The Model A didn't have a windshield, so Ernest may have gotten a bit of extra business in dental work over the years from other drivers of the Model A.

New Thinking Made Ford Different

After the Model A, the company produced a range of models, but they were expensive compared to some of the competition at the time. For example, the six-cylinder Ford cost $2,800 USD ($70,000 today), while the Oldsmobile Runabout (a competitor) cost $650 USD ($16,500 today).

As Ford cars became more successful, Ford had to find a way to reduce the cost of production while increasing output.

The secret to low cost was in the production method. At the time, fewer than ten cars could be made per day by groups of two or three men at Ford. The small groups would finish the individual components of the car before assembling them.

With the introduction of the Model T in 1908, however, a new method of production was tried. Ford looked at other industries, such as flour mills, bakeries, breweries, and meat-packing plants, and he drew inspiration from their continuous-flow methods of production. Parts for the Model T were created in bulk and then brought into the manufacturing plant. The assembly process was broken up into eighty-four steps. Each worker would be trained on just one task within the process. Ford even hired a motion-study expert, Fredrick Taylor, to make sure the jobs were made as efficient as possible. In addition, there were machines built to stamp parts faster than human beings could.

Nothing was more revolutionary in the automobile industry than the introduction of the assembly line. In 1910, Ford opened the Highland Park manufacturing plant in Michigan. In 1914, the introduction of mechanically-driven belts that moved at a speed of 6 feet (1.8 meters) per minute accelerated the pace of production even further.

The invention of the electric starter, back in 1908, took the manual effort out of starting the Model T's. This, combined with cheaper fuel and mass production, impacted the viability of electric cars, which simply couldn't match the range of their petrol-based counterparts.

The Fruits of Efficiency

The time it took to produce a car fell from over twelve hours to ninety-three minutes. To highlight how much of a revolution this was, consider that in 1914, Ford produced 308,162 cars. This was more than all other manufacturers combined. By 1924, over ten million Model T's had been built. The price had now dropped from $850 to $260 USD ($12,100 to $3,700 USD today) per unit.

Assembly work, though monotonous, was not as bad a deal for the workers as one might think. The typical workday could be reduced from nine to eight hours due to the efficiency gains. The workers were paid $5 ($124 today) a day, double the industry standard at the time. Theoretically, factory workers could afford their very own Model T's. The workers must have felt like rock stars as they tore up the street going 45 miles (72 km) per hour).

At this point, Ford had made the automobile available to the common man.

The Model T and its mass-production techniques transformed American culture and economy. Drive-in restaurants and movies began to appear, and motor hotels (motels) became a new concept. In fact, people could now drive anywhere (the cars were designed to handle rough terrain, snow, and even river crossings), and, as a result, businesses started popping up in rural areas.

Early cars and trucks impacted the steel, rubber, and glass industry, the food industry, and the entertainment industry.

By the 1950s, the auto industry was already employing nine million Americans. There was even a job called a "Tooter." A tooter's sole purpose was to listen to car horns in the factory and make sure they were tuned to the frequencies specified by the manufacturer. Car horns usually had two tones: E-flat and G, for example.

THE DAWN OF HOME ENTERTAINMENT

While the Model T was revolutionizing life outside the home, audio entertainment was doing the same inside the home.

It's somewhat hard to believe the state of home entertainment a century ago. If we want music today, all we have to do is press a button, tap a screen, or even speak. Back then, it wasn't so simple. To enjoy music at home was to be privileged enough to have family members who played instruments. Sing-alongs were also a favorite. In fact, the phonograph's initial competitor was not another audio playback machine, but the humble piano.

Even though crude by today's standards, the Edison phonograph and the gramophone brought in a new realm of possibilities in entertainment. The phonograph ushered in the idea of "private music" for the very first time. What a revolution to listen to music whenever you liked, without going to a live show. It's not difficult to see how this was a great idea for its day.

As the sound quality of records improved, they began to sell well. Along with this came mass production and effective marketing. Soon record prices dropped from two dollars to around 35 cents. In 1907, a recording of Enrico Caruso's cover of "Vesti la Giubba" became the first to sell a million copies. It was distributed by Victor Records.

GEORGE EASTMAN (KODAK): PHOTOGRAPHY BECOMES COMMON

With the press of a button or the light tap of our fingers against a glass surface, we have the power to freeze time, capturing a moment forever. A frozen image can act as a direct time portal, allowing us to look back into moments in history as they were. Prior to photography, this could only be done by a skilled artist.

Around 1900, the photographic camera was vastly different from today's. It was bulky and complicated, and photographed subjects would have to stay perfectly still for six to eight seconds to take a single picture. If they moved, the image would blur. Imagine taking a photo with your friends under these conditions. A lost cause would be an understatement.

To actually create a photo, a photographer would have to pour and mix layers of chemicals over a large glass plate to make it light-sensitive. The plate would then have to be inserted into the back of the camera lens. The picture had to be taken before the plate dried, giving it the name "wet plate photography." The problem was that the glass plates were heavy, fragile, and cumbersome to use. Only a professional photographer could put up with using such a system.

A young man by the name of George Eastman, of New York, had a casual interest in photography but struggled greatly with cameras. Even after professional lessons, he still battled with the uncooperative, messy chemicals. He knew there had to be a better way.

Eastman did the logical thing and researched to see if there were any better techniques of photography. Each night, he would study international photography journals. This turned out to be a good move, as it led to the uncovering of a new, simpler kind of photography. In England, British photographers were experimenting with so-called "dry plate" photography. These plates were pre-coated with the chemicals, so the mess was eliminated. It was a hidden gem. There was no internet back then to spread this new trend in photography to America, so very few Americans knew that the British had invented a superior photographic method.

Wet Plate Photography

Eastman then thought of the next logical step—to create his own dry plates. Why not? Being a high school dropout and never having studied chemistry could have been two reasons not to give it a go, but that didn't matter to him. He would soon get to work. Eastman had a day job as a bookkeeper at a bank. After work, he would spend all night attempting to make dry plates in his mother's kitchen, pouring chemicals onto glass plates and baking them in the oven. Eastman worked so hard that his mother would often find him asleep on the floor in the morning.

Eastman would test out his new plate technology by taking photos of a neighbor's house. One day, a local photographer saw the pictures and was curious. The local photographer bought some dry plates and recommended them to the Anthony Company, a leader in photo supplies in New York. Anthony Company was in the business of finding new and innovative products, exactly the kind of stuff that Eastman had to offer.

Anthony Company offered the twenty-six-year-old Eastman a contract. By January of 1881, Eastman had his own dry plate factory in New York, while still working his bank job. He mixed all the chemicals himself and worked late into the night. When he went to work at the bank the next morning, his fingers would often still be stained black from the chemicals. He was determined to succeed. Soon, Eastman could afford to employ a small staff.

After a disagreement over a promotion, Eastman left the bank, a move which his co-workers saw as a big mistake, but it was Eastman who would soon be laughing. Photographers quickly found out about Eastman's invention and gladly left the drudgery of wet plate photography in favor of his simple, easy-to-use dry plates.

Unfortunately for Eastman, other companies soon caught on and began selling their own versions of dry plates. Eventually, the other companies' products exceeded the capabilities of Eastman's company, and he found himself looking at failure.

Eastman had to think quickly. Soon he came up with chemically-coated, rollable paper film. Instead of replacing a dry glass plate with another one to take another photo, you simply had to wind the roll to get to the next frame of film. It was a forward-thinking idea, and Eastman was excited about it. "We'll be able to scoop the world in the next few weeks!" he exclaimed to his staff. In 1885, he demonstrated his invention. Although it won a few awards, film material limitations meant that the image quality was not up to the standards of professional photographers. As a result of the quality issues, the paper-roll film was a failure on the market. Eastman was devastated, but, in a stroke of genius, he decided to turn his attention to why he had started all of this in the first place. It was to make photography so simple that *anyone* could do it. Maybe the professionals couldn't see the use of rolls of paper as film, but perhaps the everyday individual would love to take casual snapshots, even if the quality wasn't perfect. It was worth a shot.

Bringing Photography to the Masses

In 1886, Eastman pivoted his plan from film production toward making an easy-to-use camera. It was to be the simplest possible design. Within a month of the idea's genesis, a camera was developed. Its name? The Kodak.

The Kodak was small. It had minimal moving parts, no focus and no viewfinder, and it came with the slogan, "You push the button, we do the rest." It was a revolution. No more complicated chemicals or glass plates, just a small box. The compact camera drew such interest, it was even mentioned in the novel *Dracula*, which was written in the same year. The Kodak camera was in some ways a success, but at $25 (three months' wages in 1887), it was too expensive

Brownie Camera

for most. To be a runaway success, Kodak just had to find out how to make the camera cheaper.

In 1888, George Eastman left his arrangement with Anthony Company and founded the Eastman-Kodak Company (now Kodak).

Real consumer success came with the Brownie camera. Eastman asked Frank Brownell, the company's camera designer, to design the least expensive camera possible, while at the same time making it practical and reliable. His strategy was to aim the product at children who loved taking photos. Released in 1901, the Brownie was small, cheap (two dollars for the camera and one dollar for a roll of film), easy to use, and could take eight shots on one roll. It was an instant smash hit, with 250,000 units sold in the first year.

The Brownie changed the human experience. For the first time, anyone could capture moments in their own lives, something that had never been possible before in human history.

There were many times when George Eastman could have given up, but his determination truly changed the world. Today there are billions of cameras worldwide that fit in our pockets. Although they work by different methods, they give the same result—a way to preserve the special moments in our lives.

Interestingly, as we're about to see, it was Eastman who would indirectly aid in the discovery of plastic.

Where Did Plastic Come From?

Take a look around you. How many objects can you spot that are made of plastic? Who do we have to thank for this material that is so integral today?

Leo Baekeland was a Belgian-American chemist born in 1863. He was an intelligent student and had a PhD at age twenty-one. In 1889, he was appointed associate professor of chemistry at Ghent University (Belgium). In the same year, he was offered a scholarship to visit English and American universities. A professor at Columbia University in New York City convinced the young Belgian to remain in the United States.

Baekeland had a keen interest in photography, and after two years of hard work, he invented Velox, a photographic paper that could be developed under artificial light. It was the first commercially successful photographic paper (much more so than Eastman's original paper film). In 1899, Baekeland would sell the Velox paper and the company he had formed to George Eastman of Kodak for $1 million USD ($30 million USD today).

With the money from Kodak, Baekeland was now a wealthy man and could afford to spend his time doing what he loved. In his case, that was performing chemical experiments. He built his own personal lab and made it his mission to find another area in the chemical sciences that could use his problem-solving talent.

He wrote in his journal: "In comfortable financial circumstances, a free man, ready to devote myself again to my favorite studies...I enjoyed for several years that great blessing, the luxury of not being interrupted in one's favorite work."

Many chemists in the late 1800s were trying to invent a new type of synthetic material that was hard and resistant to heat. Organic resins such as tree rubber existed but were not tough enough to have a wide array of applications. Chemists were using two chemicals to try to form this new synthetic material: phenol (from coal and tar) and formaldehyde (from the combustion of methane).

A point of note: Formaldehyde is now known to cause cancer and can be found in car exhaust fumes and tobacco smoke. The experiments of these early chemists were dangerous without them even knowing it.

One such chemist, Adolf von Baeyer, succeeded only to create a "black guck" which he called "useless." A student of Baeyer's created a "mess" which he couldn't "do anything with once produced."

Nevertheless, Baekeland aimed his research and investigations at phenol and formaldehyde, hoping to succeed where others had failed. Painstakingly, he studied the experiments of Baeyer and others, slowly changing the variables until something useful emerged. By individually modifying the temperature, pressure, and amounts of each chemical used, he eventually struck gold...or plastic. He found a material that was strong, lightweight, cheap, didn't shatter or crack, was resistant to heat, and best of all, could be molded into virtually any shape. He called this new material Bakelite, and it was the very first plastic. In 1909, he patented his invention and announced it to the American Chemical Society.

It was an instant success. This new material also took colored dyes well, which could make for bright and attractive products. Early uses included buttons, radios, telephone cases, lamps, chess sets, billiard balls, and plastic toys. Baekeland would receive many awards for his invention, and today Baekeland plastic products are considered a rare item for historical collectors.

THE WRIGHT BROTHERS DEFEAT GRAVITY

By 1900, becoming airborne was by no means a new achievement. The French had been flying in their air balloons since 1783. In 1900, the first Zeppelin flew in Germany. However, floating, as these machines were doing, is a very different thing from flying like a bird in the way great minds like Leonardo da Vinci envisioned.

The first machines built for serious flying attempts were more than bird-like inspirations. They were often a direct copy, with large, mechanical, flapping wings (these machines are known as ornithopters).

Early flying was a very risky business. If you had an incorrect assumption about how a wing works, for example, it could lead to an early death.

Franz Reichelt was a tailor who believed a wearable parachute was a step toward human flight. After some successful initial experiments involving dropping the parachute with test dummies from the fifth floor of his apartment building, Reichelt went on to test the technology on himself. In 1912, despite opposition from his friends and family, Reichelt jumped from the Eiffel Tower with his wearable parachute. The parachute failed to deploy, and he was killed instantly at the point of impact.

Otto Lilienthal made over two thousand glides in a hang glider in Berlin. His only form of control was to shift his body weight. Unfortunately, his glider plummeted to the ground after stalling during one of his test flights. He survived the initial crash but died from his injuries thirty-six hours later. Lilienthal's final words were "Sacrifices must be made." These words were also engraved on his tombstone, a somber reminder of what the mastery of flight meant to those brave enough to try.

Others had more success.

French inventor Clément Ader was the first to achieve self-propelled flight. The machine mimicked a flying bat. His first flight covered 50 meters on October 9, 1890. Later reports say he flew 180 meters in 1892.

Gustave Whitehead, a German immigrant to the United States, flew a steam-powered plane in 1899.

By now, many uncontrolled glides and short powered flights had been made. What had not been achieved was a heavier-than-air controllable craft that could carry a person under its own power. Many engineers and scientists around the world were trying to solve the mystery of flight, then known as the "Aerial Navigation Problem." The year 1900 had witnessed a whirlwind of innovation in transport, and there

was keen public interest. The American and European people were
thinking: after all, automobiles were here, electric cars were here...
surely a flying machine had to be next!

In America, the humble bicycle was also taking over in a big way.
In 1892, two brothers, Orville and Wilbur Wright (twenty-five and
twenty-one years old, respectively) decided to get into the bicycle
business. They started up a small bicycle shop in Dayton, Ohio, where
they manufactured products by hand. After five years of this, however,
Orville grew restless and felt he wasn't using all of his talents. Wilbur,
meanwhile, had been watching the aviation progress from a distance.
He decided that the brothers should give controlled flight a shot.

Control was the hardest part of flying. How do you steer something
that is already in the air? While thinking about the problem,
Wilber was standing outside observing birds in flight. From careful
observation, he thought that maybe birds managed to turn by slightly
twisting their wings. The brothers made a small box kite to test their
idea. The string was attached to the wings in such a way that, when
pulled from either end, the string would twist the wing structure one
way or another. This was called wing warping and was a simple but
substantial breakthrough in aviation control.

The test kite worked perfectly, and the kite turned as expected
when the wings were pulled and warped. In 1900, the brothers
went on to build a full-scale version with additions to the control
mechanisms. They tested this full-scale model at Kitty Hawk—a
beach in North Carolina.

They made several successful small glides down a hill, but disaster
struck in 1902. One day, as Wilbur was gliding and performing a
turn, his body began sliding down the side of the wing, and he quickly
lost control. The aircraft slammed into the ground. He was violently
thrown and hit his head on one of the aircraft's support struts. Shaken
but alive, a discouraged Wilbur admitted that man would not fly for
another fifty years.

The brothers didn't give up, though. They resumed aviation
experiments at their bike shop a few months later. In a flash of
genius, Orville decided to build a small-scale wind tunnel in the shop,
to validate their equations and test different wing shapes. The wind
tunnel was a small wooden box with a 30-horsepower fan at one end
and a glass top for observing the wings that were placed inside.

Orville measured each wing's performance in detail until he found the best design. Incredibly, this entire process of testing was pretty close to the modern engineering method. Impressive, especially considering Orville was a bike mechanic from 1902 with three years of high school education.

With the wind tunnel set up, Orville was the first to realize that the behavior of a wing is the same regardless of size—it was only the shape that mattered. Because of this fact, the brothers no longer had to risk their lives by building full-size models. There was no need to jump off mounds to see if a wing design worked or not.

By 1902, they had perfected their design and could now reliably control the glider.

The next year, they were ready for the main show—powered flight. Using the skills of an in-house machinist from their bicycle shop, the brothers produced an engine and a pair of propellers.

However, things were about to change. Some details about their glider had been leaked and published in a French journal, re-sparking an international interest in flying. The Wright brothers had a head start, but it wasn't for long. The French were also turning their eyes to the skies. The race to conquer the air was on!

Taking to the Skies

It was the morning of December 17, 1903, at the sandy beaches of Kitty Hawk. The engines and propellers had been fitted to the aircraft, and it was ready for its first secret test flight. The brothers shook hands somewhat solemnly, as though it was their final time seeing each other, and split—Orville to the plane, Wilbur to observe. Orville started the engines. A loud noise could be heard as the engines began to rev. History was about to change.

Just seconds later, the plane steadily moved forward and lifted off the ground. It flew for twelve seconds. The world's first fully-controlled powered flight had been achieved. They flew four times that day, the longest flight being fifty-seven seconds. Man had taken to the air. The brothers telegrammed the details to their father in Ohio.

By 1905, with further improvements, the Wright brothers were flying for thirty minutes. However, the brothers had an issue. They were looking to sell their invention, but how? They were afraid that, if they demonstrated the technology publicly, it would be stolen, as they had not yet obtained a patent. They were in the unusual position of trying to market their "flying machine" without ever showing it to a customer. Regardless of this fact, their first thought was to target the US government as a potential customer. The deal went something like this: Sign the contract and give us a list of things you would like our flying machine to do. We'll go out and make sure it can do those things; then, if it does, you can pay us.

The American government rejected the offer.

The brothers then turned to Europe. At this time, Germany and France were on the verge of war, so the pair thought that either party would be very interested in just such a flying machine. The asking price? $200,000 USD ($5.2 million today). Both countries refused to pay so much for something they couldn't see.

Meanwhile, in Paris, powered flight had just been achieved by Alberto Santos Dumont. The French were catching up. Despite the plane only being able to fly in a straight line, the people of France were amazed and hailed Dumont as the father of aviation.

Back in America, the Wright brothers had received their patent (for aerodynamic control technology), but they still wouldn't show anyone the plane. They feared their design secrets would be stolen by spying onlookers. With no physical plane and only claims, doubts began to rise over the legitimacy of the brothers' assertions. The French were especially scornful and suspicious; they were fast becoming the perceived leaders in aviation, while the Wright brothers sat on their invention.

By January 1908, Frenchman Henri Farman was flying in a circle over a total distance of 1 kilometer. The control was crude, but it was control nonetheless. The Wright brothers were starting to look like fools on the world stage.

Fortunately, word reached President Roosevelt of the claims by the Wright brothers. He sent the Army to investigate this so-called "flying machine." The Army negotiated a $25,000 deal with the Wright brothers to supply them a plane.

The deal was as follows: Produce a plane capable of carrying two men 125 miles (200 km) at 40 miles (64 km) per hour for a flight time of one hour.

The brothers agreed. Now, they could finally prove to the world that their machine flew.

Their first targets were the skeptical Europeans, particularly the French. On August 8, 1908, a crowd gathered near Le Mans in France to witness the Wright brothers' flying machine with their own eyes. In the middle was the Wright brothers' plane with Wilbur at the controls. With all eyes focused on him, he started up the engine and was flying a short time later. The crowd rose to its feet in simultaneous astonishment.

As the plane made smooth, controlled turns, it was clear that the brothers had not been lying. They had been vindicated! The French news outlets admitted that this aircraft was far beyond what their local engineers had achieved. "We are beaten," remarked some of the French papers of the day.

Today, there are over 100,000 flights per day and over 37 million every year. Of course, there were many contributions to aviation after the Wright brothers, but the fact remains that the true origins of air travel were very humble: two bicycle makers with a high school education and a keen sense of intuition.

Neil Armstrong (also from Ohio) carried a piece of the Wright brothers' plane with him as he set foot on the moon in 1969. It was a piece of fabric from the left wing.

And so, it went: By the year 1910, humanity had mass-produced cars, personal music, aircraft, small personal cameras, and plastic. The opening decade was a sign that this century would be very different from the last.

CHAPTER 4

Innovation and the Great Conflict 1910–1919

Carrying on from the last decade, the 1910s continue with innovation and the expansion of new ideas. In 1912, a complete concept of "continental drift" (movement of land masses) is introduced by German geophysicist Alfred Wegener. Also, in 1912, the Titanic sinks, and the first Olympic games with electronic timing (and photo-finish) technology take place in Stockholm, Sweden. The record for the 100-meter sprint is set by Donald Lippincott with a time of 10.6 seconds.

Stainless Steel

The year 1912 was also the year stainless steel was invented by Harry Brearley of Sheffield, England. In the build-up to World War I, the UK was strengthening their military might. The Royal Arms Munitions factory in London was having some difficulty with erosion inside their rifle barrels, causing poor accuracy. They invited Brearley, who was head of Brown Firth Labs (a steel maker), to take a look at the situation. After experimenting with adding more chromium to a mix of metals used to make standard steel, Sheffield stumbled across something peculiar—a steel that did not corrode or rust. A stainless steel.

At the time, steel rusting over time was a fact of life—as sure as wood rotting. Surprisingly, Harry was virtually alone in his vision of the endless applications for stainless steel. His superiors thought further applications were a waste of time and that he should leave his tinkering alone. Following World War I, the full potential of stainless steel was finally realized. Today, stainless steel is used in everything from cars and cutlery to buildings and electronics. Currently, almost 46 million metric tons of steel are produced globally.

First Animated Feature Film

The 1917 seventy-minute film, *El Apostol*, by Quirino Cristiani of Argentina was the first animated film ever made. It was well received by audiences. The film was a tongue-in-cheek commentary on the politics of Argentina. It featured the president ascending into the clouds before using thunderbolts to rid Buenos Aires of all corruption, leaving behind a burning, smoldering city. The film would literally be smoldering in 1926 when the film studio it was in was consumed by a fire. There was only one copy, meaning it was unfortunately lost forever.

Toast, Anyone?

Long before millennials and their obsession with avocado toast, the first toasters prepared bread in front of a hot fire. The first mass-produced toaster was the "D-12," manufactured by General Electric and invented by Frank Shailor in 1909. It was a bit of a pain because you had to stand next to the toaster and turn it off when the toast looked done. It also only did one side at a time.

The year 1914 saw a Westinghouse version of the toaster, but it wasn't much better.

The real breakthrough came in 1919, when Charles Strite invented the first pop-up toaster, which used heated electrical coils to toast bread. The problem back then was that all bread was cut by hand, so slices were of different thicknesses, but over the next ten years, bread-slicing machines gained in popularity. As this happened, slice sizes became standardized, enabling the widespread adoption of the electric pop-up toaster. Today, the toaster is the most common household appliance in the United States.

WORLD WAR AND INVENTION

While innovation was moving along at a steady pace, things were beginning to heat up politically. The world would soon erupt in the worst war yet seen. As you'll see later in the chapter, it was against this backdrop of the horror of war that rapid technological advancements were made. To fully understand the scope of technological progress at that time, we must understand the events surrounding the First World War.

Battle Beginnings: The Powder Keg of War

The world's power balances were changing. In 1910, the eight-hundred-year-long monarchy ended in Portugal. In 1911, Italy and Turkey were at war. A 1912 revolution overthrew China's last dynasty to create the Republic of China. All this while, alliances had slowly been forming in Europe. On one side were Great Britain, France and Russia (the Triple Entente), and on the other, Italy, Germany and Austria-Hungary (the Triple Alliance). The situation was a powder keg ready to explode.

With the industrialization of most of Western Europe, and the newly formed alliances, an arms race was underway. A freshly emergent and unified Germany baffled Europe with its prosperity and technical abilities. Great Britain and France in particular were increasingly worried about this new German nation. Meanwhile, the second largest country in Europe, the fifty-year-old Austria-Hungary, was faced with growing dissatisfaction from its Serbian Slavic community, who made up a staggering 50 percent of the population. The Slavic (ethnic eastern European) people felt they were not treated as equals with the Austrians and Hungarians in the region.

The Austro-Hungarian empire was an industry powerhouse. It had the fourth-largest machine-building industry on the planet, behind the United States, Germany, and the UK. Austria-Hungary was also the world's third-largest manufacturer and exporter of electric home appliances, electric industrial appliances, and power-generating equipment for power plants, behind the United States and the new German empire. They were a technologically advanced nation for the day and a power to be reckoned with.

The leader of Austria-Hungary was eighty-one-year-old Franz Joseph I, who was unwilling to change his ways of governing, which involved marginalizing the Slavs. The successor to the throne was Archduke Franz Ferdinand. He was a radical departure from Joseph I and wanted to bring peace to the nation by giving the Slavs a fair chance.

June 18, 1914: While on a tour of Sarajevo, Bosnia, Ferdinand is greeted warmly by the people. That is, until a bomb bounces off his car and explodes behind him. Despite suggestions to cancel his tour, he continues down the road, only to be shot and killed by Gavrilo Princip, a nineteen-year-old Bosnian radical. Princip had conspired with five Serbs on the assassination but failed to understand that

Ferdinand wasn't the bad guy and was, in fact, in favor of the equality of the Slavs. The goal of the assassination was for the southern Slavic states to break free of Austria-Hungary and form their own state: Yugoslavia.

It wouldn't be long before the eighty-one-year-old leader of the Austro-Hungarian Empire declared war on Serbia as revenge (they were considered brothers of the Slavs).

This declaration was all it took for the global alliances to click into place: The Russians, British, and French backed the Serbians, while the Germans and Ottoman Turks backed the Austro-Hungarians. This single bullet that killed Franz Ferdinand was the match that ignited the powder keg.

WORLD WAR I: THE GREAT CONFLICT

At the start of the war, armies were riding horses. By the end, they were fighting with tanks on the ground and planes in the air. When your country is at stake, you must innovate to win or die trying. These factors led to World War I being the first industrial and scientific war.

Although Austria-Hungary was technologically advanced, they were falling behind in the arms race due to economic troubles. For example, the country was still attempting to use balloons for reconnaissance missions instead of planes.

The Germans and Austro-Hungarians thought the battle with Serbia would only last four to five weeks, but the Serbian army (one-tenth their size) put up an incredible fight and actually defeated Austria-Hungary in three battles with great losses. The conflict was quickly shaping up to be more difficult than first thought.

Meanwhile, Britain, the economic and military superpower of the world and undisputed leader of the sea, had a plan to bring down Germany. Britain used its shipping might to block off food supplies to Germany to starve the nation into submission. In retaliation, the Germans used their ships to block off food and military supplies to Britain. The Brits had somewhat expected this, but they were not prepared for a new innovation the Germans had up their sleeve: a brand new weapon named the Unterseebot (meaning "undersea boat"), or U-boat. A U-boat was basically a submarine. Although common today, this required an immense amount of engineering expertise with the technology available in the 1910s to reliably work in combat. The German U-boats could launch torpedoes at enemy ships without them ever seeing their destruction coming.

Innovation at sea would only be outdone by innovation in the air.

Advances in the Air

By 1915, the world's first strategic air attack was conceived by the Germans, who bombed Britain using blimps.

Not long after, the Germans gained air superiority on the Western front by designing one of the first planes able to fire a machine gun through its propellers. They did this with a mechanical device called a synchronization gear.

By 1917, with the ship blockades still in effect, the citizens of most of the European powers were beginning to suffer food shortages; the war needed to be decisively won. As a result, aeronautical engineering—which began with the Wright brothers just a short time earlier—was advancing at a rapid pace. And soon, fierce dogfights were happening in the sky.

Britain's answer to Germany's air power was strategic bombing. Large British biplane bombers were soon built; these planes were some of the largest aircraft in the world at the time.

Bentley and BMW's First Gig Was in the Sky

Before the war, Walter Owen Bentley and his brother, Horace Millner Bentley, were car salesmen in London. Deep down, however, Walter had always wanted to design and build his own cars, not sell other people's. One day, in 1913, he noticed an aluminum paperweight and an idea was sparked. What about replacing the heavy cast-iron pistons in cars with lighter aluminum ones? During the war, these lightweight Bentley aluminum pistons would soon be used in the famous British Sopwith Camel aircraft engines. After the war, Walter used his experience to found Bentley Motors Ltd. Today, Bentley is arguably the most prestigious car brand on the planet.

In April 1917, Bayerische Motoren Werke (BMW) came into existence. BMW's first product was the BMW IIIa aircraft engine for the German war effort. The IIIa engines were good performers and fuel-efficient. The war effort meant that BMW could expand its operations quickly, if only in aeronautics. This first engine demolished the competition. Planes equipped with it could outmaneuver and outclimb anything else in the air. Along with speed records, this BMW engine helped power a biplane to a record 32,000 feet (9,750 km) in 1919. That's roughly the altitude you cruise at today in a passenger jet, and that was in 1919!

Today, BMW still designs the interiors of the first-class sections in Singapore Airlines jets.

The Birth of the Documentary

As the war raged, there was great interest in the events. Newspapers could tell the story, but the newly formed film industry could do a much better job. A picture may be worth a thousand words, but fifteen to thirty pictures every second can be worth much more. Some films of this period took on a different structure. Their sole purpose was to document events. These types of films would be known as documentaries, and the war helped the format come to prominence.

An End to Conflict?

It was the lower ranks of the Russian army that rose first against their superiors. The army had lost 1.8 million men and enough was enough. The Russian civilians were starving and there was revolution in the country. The nation pulled out of the war in November 1917.

In Russia, the war had caused economic destruction, allowing the perfect backdrop for one of the biggest revolutions the world had ever seen: the establishment of Communism.

Exhausted, the French army also refused to fight, and the nation pulled out soon after. Now it seemed the Germans and Austro-Hungarians had the upper hand.

It wasn't until a German U-boat sank a passenger liner carrying Americans into British waters, and the Germans prompted Mexico to attack the United States, that the US was forced to enter the war.

Fresh American troops helped push Germany back and turn the tide. The final assault from Germany wasn't enough and, defeated, they signed an armistice agreement on November 11, 1918. When the dust finally settled, a total of 37 million military personnel and civilians had lost their lives.

The war's end triggered the collapse of five of the last modern empires: Russian, Chinese (due to internal turmoil), Ottoman (Turkey), German, and Austro-Hungarian. Another era had ended, and a new one had just begun.

PERSEVERING THROUGH TOUGH TIMES

Although war had ravaged the world, at this time, there were several creative and dedicated individuals building companies which would become household names. In some cases, the demands and wider ramifications of the war actually served as a springboard for their success. We'll visit a few of those stories now.

Sharp Corporation: It all Started with a Pencil

In 1915, Tokuji Hayakawa, a Japanese businessman, received a contract to make metal fittings for use in a mechanical pencil. After studying the design of the entire pencil, he noticed that it was fragile, and he thought he could do better. Seeing the business potential in a practical mechanical pencil, Hayakawa got to work, often neglecting eating or sleeping. Eventually, he managed to improve the internal design, making it much more durable.

With more tweaking, he made the device easy to use and reliable. The result was a modern, twist-type, mechanical pencil (called "Ever-Sharp"). Tokuji Hayakawa would team up with his brother Masaharu Hayakawa to sell the pencils, producing 1,400 a month.

To get the ball rolling, the pair went to a number of stores to try to sell their product, but the stores were critical and showed little interest. However, the brothers didn't give up. They eventually came across a high-class stationery store that showed interest, though the store requested thirty-six revisions of the product from the brothers before finally accepting it.

Strangely, it would be World War I that was their savior. The breakout of war in 1914 made it difficult for the US and Europe to get hold of German-made mechanical pencils. The brothers' pencils were the only ones that matched the quality of the German product. The pencils began to gain popularity in the US and Europe as a consequence. This Ever-Sharp mechanical pencil is where the Sharp Corporation derived its name. From pencils, Sharp went on to would create Japan's first mass-produced microwaves and solar cells, as well as the world's first all-transistor desktop calculator.

Sharp's innovations continued throughout the decades and today, under the ownership of Foxconn, they have 42,000 employees and $15.1 billion in assets.

Panasonic: Picking a Gap in the Market

Panasonic (originally "Matsushita Electric Housewares Manufacturing Works") was founded in 1918 by Konosuke Matsushita. Matsushita was the son of a wealthy businessman who lost all his money in rice speculation.

In 1918, Konosuke Matsushita set up Panasonic in his two-story home in Osaka, Japan. The staff consisted of three young people: the twenty-three-year-old Matsushita, his wife, twenty-two-year-old Mumeno Iue, and Mumeno's fifteen-year-old brother, Toshio Iue. Interestingly, Toshio would go on to found Sanyo.

Although Panasonic's first product was a fan insulator plate, Matsushita had great insight into the world of business and was convinced there was a huge untapped market for well-designed, high-quality household electrical fixtures. Matsushita saw a gap in the market and went straight for it.

Matsushita stayed up late at night refining his designs, ultimately choosing to manufacture two new products: an attachment plug for light bulbs, and a two-way socket (also for light bulbs). They proved popular, as they were of higher quality than other products on the market. With a small staff, Matsushita could keep his costs down. As a result, his products were 30 to 50 percent cheaper than the competition's.

By the end of 1918, the company was employing twenty people. In the coming century, Panasonic would move on to bicycle lamps, then electrical components such as vacuum tubes, electric irons, radios, record players, VHS VCRs, and even a Japanese-made IBM PC in the early 1980s.

Today Panasonic has 257,533 employees, and $65 billion in revenue, $53.3 billion in assets.

IBM

In 1911, four companies (The British Tabulating Machine Company, International Time Recording Company, Computing Scale Company, and the Bundy Manufacturing Company) merged to form the Computing Tabulating Recording Company (CTR) in New York.

CTR manufactured a wide range of products, including coffee grinders, automatic meat slicers, employee time-keeping systems, weighing scales, and punch-card equipment. Although CTR's company make-up didn't change, the name was changed to International Business Machines (IBM) in 1924.

It was the punch-card business that would prove to be IBM's strongest business at its genesis.

Back then, a computer was a person who calculated arithmetic sums. It was an actual profession—not hard to see where the modern computing machine first got its name. We will explore IBM further in the 1980s (chapter 11) when we look at computing.

Today, IBM has 380,000 employees and is an emerging player in machine learning.

Boeing: Crash My Plane and I'll Build a Better One

It had only been thirteen years since Orville and Wilbur Wright first took powered and controlled flight, but an industry in the air was already forming. William E. Boeing would become one of the biggest names in the industry. Boeing had a background different from what you might expect. He was a lumber company executive from Michigan. William Boeing had success shipping lumber to the East Coast of the US using the newly established Panama Canal.

In 1909, Boeing saw his first plane at an exposition in Seattle and was fascinated. He took flying lessons and soon bought his own plane, though it was an accident that created one of the largest aircraft manufacturers the world would ever see.

Boeing's test pilot, Herb Munter, damaged Boeing's plane during a flight. Upon hearing that the replacement parts wouldn't come for months, Boeing remarked to his friend Conrad Westervelt of the Navy, "We could build a better plane ourselves and build it faster."

Boeing asked Westervelt to design a plane, which Boeing would build. The result was an amphibian seaplane with impressive performance. Confident in their abilities, the pair decided to start their own company, Pacific Aero Products Co., in 1916.

In 1917, America entered World War I and the nation needed planes. The US Navy signed a contract with Pacific Aero Products Co. (who had just changed their name to Boeing Airplane Company) for fifty aircraft.

When World War I ended in 1918, a surplus of cheap, used military planes flooded the commercial airplane market, preventing aircraft companies from selling any new planes and driving many out of business. Boeing was forced to start manufacturing other products. Strangely, Boeing would build dressers, counters, and other furniture, just to stay afloat in this tough time!

Gradually, as the market improved, Boeing went back to aircraft, concentrating on commercial aircraft this time. Boeing started with airmail carriers, which later led to passenger services. One of these ventures would directly result in United Airlines.

Today, Boeing has 147,680 employees making 748 commercial aircraft, 180 military aircraft, and around five satellites per year.

The Boeing Factory

The current Boeing Everett factory in Washington is so large that, when it was being built, there were clouds forming near its ceiling. The factory has over one million light bulbs, and there are 3.7 km (2.33 miles) of pedestrian tunnels running beneath it.

End of a Decade

By the end of the 1910s, the Great War had accelerated aircraft technology to what can be seen as the beginnings of the modern era.

The world had been reshaped: People were tired of hardship and ready to move forward. A time of peace and technological advances followed, bringing unprecedented economic growth for most of the world. In the United States, this period would be affectionately known as the *Roaring Twenties*, and more quaintly in France as the *Crazy Times*.

CHAPTER 5

Gears in Motion
1920–1929

When researching this chapter, I was surprised to see just how many mainstream inventions came from the 1920s. While the automobile was changing culture, the 1920s was also the beginning of mass adoption of electronic products like telephones, radio, and kitchen appliances. Instant media broadcasts also changed culture for the first time. Celebrities were no longer scientists and philosophers, but sports stars and movie stars. Other novel inventions of the 1920s include the children's jungle gym, bubble gum, hair dryers, the car headrest, the lie detector, bulldozers, cheeseburgers, and the drive-through (oddly, first invented by City Centre Bank in 1926).

In this chapter, you'll see the technological bedrock that was being laid for the twentieth century. TV, color movies with sound, mass production of consumer goods—these are all things that didn't exist in the 1910s.

THE MOST PROSPEROUS NATION ON EARTH

There was significant economic prosperity in the United States following World War I. Mass production through mechanization was the name of the game. Higher profits were generated as new machinery and manufacturing methods were introduced, and an increase in living standards followed. At the same time, automation broke out from the automobile industry and became mainstream—everything from biscuits to fridges was now mass-produced. Life seemed much easier, not only in contrast to the horrors of war, but also because technology had become part of people's lives. This technology brought jobs: boilermakers, riveters, foundry men. The lingering grasp of steam power was finally set aside as it gave way to oil and electricity.

By the end of the 1920s, a lot of the middle class in the United States could enjoy a lifestyle never seen before. It was no more evident than in the "modern" home. Technology became a part of everyday life, and the middle class had the extra money to afford it. The vacuum cleaner replaced the carpet beater. Mass-produced electric fridges, sewing machines, and ovens now existed. Everything from the pop-up toaster to the food blender saved countless hours of work in the home.

Mass production also brought in the age of mass advertising, and $1 billion per year ($14 billion today) was spent on advertising in the United States. These were good times, with new forms of art and music seeping into culture. In Europe, however, things were very different.

Germany After World War I

The Treaty of Versailles (a forced admission of guilt, and a set of economic and political punishments against Germany) destroyed the German economy. Out of 440 clauses in the agreement, 414 were aimed at punishing Germany. In 1922, German inflation was out of control, and goods were rapidly getting more expensive.

Employers would pay citizens with suitcases full of money. As soon as they were paid, employees would have to literally run to the store. By the time they got there, the prices would have risen. Prices doubled every two days at the peak of inflation. To give you an idea, in January of 1923, a slice of bread cost 250 marks. By November of the same year, it cost 200 million marks.

The Rise of the Car

One of the biggest drivers of the American manufacturing economy was the automotive industry. Henry Ford's impact on mass production created jobs and a cheap enough product for the working class to afford. In this way, America was able to ride out its initial halt in prosperity in 1920.

Sales of the Model T hit ten million in 1924. The car became one of the biggest businesses in the US. Even car scrapyards boomed. With employment at an all-time high, factory workers were even buying shares in the stock market. By the mid-1920s, one in five Americans owned a car. It would take Britain forty years to match this rate of car ownership.

America was now the leader in manufacturing, and Henry Ford's production methods were spreading all over the nation, even to

Europe. Fiat in Italy was one of the first in Europe to copy Ford. Interestingly, their factory had a test track on the roof!

RADIO CHANGES CULTURE

Today, radio seems like a relic of the past, but this technology had a huge impact on popular music, advertising, and culture as a whole. A new technology that allowed anyone to hear information about the world rapidly—much faster than any other medium—was a big deal. In 1920, radio exploded. There were five thousand home listeners at the start of the year, and eighteen months later, three million families had radio. Four years later, this number would be fifty million families.

In the 1920s, everybody tuned into the radio, and it became the very first form of "instant mass media." People from across a nation could have an instant shared experience. In the evenings, families gathered around the radio, and seeing a crowd on the street listening to the latest news was not unusual. Broadcast plays, sports, and music were a great delight for the people of the day.

Interestingly, the word "broadcasting" was originally an agricultural term for the wide scattering of seeds. It does make sense when you think about it.

Eat Your Wheaties

Advertisers soon also took advantage of this new medium. On Christmas Eve of 1926, the cereal company Wheaties aired the first commercial jingle, and immediately became the best-selling cereal in the United States. The potential of radio was immense, and by 1941, two-thirds of radio programs carried advertising.

Radio's Lasting Impact

Television owes a significant debt to the Golden Age of Radio. Major radio networks such as NBC, ABC, and CBS are still powerhouses within the media industry today. Entire musical genres, such as jazz and rock and roll, were enabled by radio. Perhaps one of the most respected companies of the twentieth century that arose from radio was the BBC (British Broadcasting Corporation).

In 1920, the first British radio broadcast was made. The public were enthralled, but broadcasts were soon banned, as it was thought to interfere with military communications. The public spoke out, and by 1922 there were petitions and many license requests to broadcast. The licensing authority of Britain decided to issue only one license, to a single company owned by radio receiver manufacturers. This eventually became the BBC. Interestingly, it was originally financed by the sale of radio sets from the manufacturer members.

War of the Worlds Broadcast

There is perhaps no greater illustration of radio's power than the infamous War of the Worlds CBS broadcast from New York. On Halloween night in 1938, a radio producer named Orson Welles told listeners that they would be listening to an audio adaptation of the science fiction novel on alien invasion, *The War of the Worlds*. Sounds like a normal evening presentation, but there was one problem. Some listeners tuned in late and missed the disclaimer that the broadcast was just a performance. The latecomers thought it was a real news story and that cities were being attacked by aliens. It sounds funny today, but back then it was truly traumatic. Almost two million people believed the story to be true. Some listeners called loved ones to say goodbye or ran into the street armed with weapons to fight off the invading Martians. Two Princeton University professors spent the night searching for the meteorite that had supposedly preceded the invasion. As calls came in to local police stations, officers explained that they were equally concerned. It eventually came to light that the broadcast was a fictional story, though I'd imagine the officers were both relieved and annoyed.

WHO INVENTED THE TV?

The predecessor of television goes back to the 1880s with the Nipkow disk by Arthur Korn, but the technology at the time failed to yield any form of device for viewing meaningful images.

Enter John Logie Baird, a Scottish engineer. In 1923, at thirty-five years of age, Baird rented a workshop in Hastings, England. He set out to solve the problem of television. With no corporate backing, he needed to get creative. Using only regular household items, such as the box packaging of a hat, a pair of scissors, some darning needles, a few bicycle-light lenses, a used tea chest, and glue, he created the first primitive television.

Two years later, in March of 1925, Baird gave the first public demonstration of his system at a department store in London. At the time, it was only capable of displaying silhouettes. On the October 2, 1925, the first successful TV transmission was made—a black-and-white image of a dummy named "Stooky Bill." This broadcast was between two rooms in his London laboratory, and the image had thirty lines of resolution at five frames per second.

As the picture came up on the display, Baird was most pleased. He stated: "The image of the dummy's head formed itself on the screen with what appeared to me an almost unbelievable clarity. I had got it! I could scarcely believe my eyes and felt myself shaking with excitement."

A dummy was nice, but he quickly realized that a human face would be more exciting. Unable to tame his eagerness, Baird got hold of office worker William Taynton. Taynton, wondering what all the fuss was about, would soon be the first person to be televised. Despite the breakthrough of being the first man on television, Taynton was less than impressed.

When Baird asked what he thought of his invention, Taynton replied, "I don't think much of it, Mr. Baird, it's very crude." Undeterred and looking for publicity, Baird visited the *Daily Express* newspaper to promote his invention.

To give you an idea of how outlandish the idea of a visual wireless broadcasting system was, a *Daily Express* news editor was terrified, and was quoted as saying to one of his staff: "For God's sake, go down to reception and get rid of a lunatic who's down there. He says he's got a machine for seeing by wireless! Watch him—he may have a razor on him."

By January of 1926, Baird's TV had been improved to show 12.5 frames per second (about half as smooth as today's broadcasts). A color version arrived in 1928, and by 1929, the first television program was produced by the newly established BBC in London, with the help of Baird's new company—the Baird Television Development Company Ltd.

Baird set up a branch of his company in France, making it the first French television company. More feats would follow, including the first televised drama show in 1930 and first outdoor broadcast in 1931, through the BBC. The days of Baird's mechanical systems were numbered, however. In 1923, Vladimir Kosma Zworkin invented the cathode ray tube, a fully electric way of doing what Baird's mechanical system did.

In the later years of the 1930s, electronic TV systems came into the picture, and Baird's mechanical system was finally abandoned in 1937. Baird wasn't done, though. He continued to work with new electronic technology, and on August 16, 1944, he gave the world's first demonstration of a practical, fully electronic color TV. Notably, Baird had also invented a 1000-line imagining system (more lines mean higher resolution). Theoretically, the image quality would be comparable to today's HDTV. Back in 1943, this was an incredible achievement! Unfortunately, plans to implement the system fell through, and its spiritual successor wouldn't be realized for another sixty years, until 2003.

Baird, a Colorful Man

Baird was a rather eccentric character, and some of his inventions mirrored this. These included a glass razor (that unsurprisingly shattered), pneumatic piston shoes (which burst in spectacular fashion), and a method of creating diamonds by heating graphite (unfortunately shorting out Glasgow's electricity supply).

One day, during one of his unusual experiments, he burned and shocked himself on a 10,000-volt supply. Fortunately, it was only his hand that was burned. Unfortunately, this resulted in an eviction from his property. Despite his oddities, Baird has been honored around the world: In Australia, the Logie Awards are named in honor of John Logie Baird.

Following a UK-wide vote in 2002, Baird was ranked number 44 in the BBC's list of the 100 Greatest Britons.

In 2014, he was inducted into the Society of Motion Picture and Television Engineers.

Film Gains a Sonic Dimension

Post-WWI, Europe's exhaustion and physical destruction made film production scarce. This opened the door for US films to take center stage. One of these films would change the world of cinema forever.

The Jazz Singer, released in 1927, was cinema's response to radio. It was the first feature-length movie with sound. The technological breakthrough for *The Jazz Singer* was enabled by something called a Vitaphone, invented by Warner Brothers and First National. The soundtrack to the movie was played on a separate phonograph record in sync with the projector motor, a feat which had not been reliably achieved up until that moment.

"Vitaphone" means "living sound" in Latin and Greek, respectively.

Hearing sound that corresponded to vision was nothing short of awe-inspiring for the audience. The Jazz Singer smashed previous box-office records and established Warner Brothers as a major Hollywood player.

AIRCRAFT BECOME THE FUTURE OF LONG-DISTANCE TRANSPORT

Airmail was trialed in the early 1920s, but it was dangerous. Of the forty pilots flying on the initial postal route, thirty-one died in crashes. Although the Wright brothers had proved that controlled flight was possible, traveling reliably by air was far too dangerous and would never become a reality...or so the public perception went. This would all change in 1927.

A twenty-four-year-old Charles Lindbergh, just one year out of army flight-training school, was one of those airmail pilots. He was delivering mail from St Louis to Chicago for $81 ($1,414 today) per week.

At the time, there was a running competition to fly solo non-stop from New York to Paris for a prize of $25,000 ($352,000 today). At the time, planes were viewed as too dangerous to complete such a feat, and six pilots had already lost their lives attempting this treacherous journey across the Atlantic.

Lindbergh believed that he had the skills to do the impossible; he dreamed of changing history. The biggest challenge of the flight was deemed to be simply staying awake. To prepare for this, Lindbergh went fifty-five hours without sleep before the flight to test his endurance.

The Pivotal Flight

On May 20, 1927, Lindbergh took off from Long Island, NY in his $10,000 custom monoplane, *The Spirit of St Louis*. He was going to attempt the challenge. Lindbergh had packed a modest meal of five ham sandwiches for the thirty-three-and-a-half-hour journey. Unfortunately, the sleep-endurance training wasn't enough. Twenty-four hours into the trip, he reported hallucinations, claiming that the "fog islands" moving below the plane were "speaking" words of wisdom to him. Despite this, Charles fought the sleep and pushed through.

At the time and place Lindbergh was scheduled to land in Paris, there was an ecstatic crowd of 150,000 waiting for him. When he finally touched down, Lindbergh became an instant celebrity, and the most famous American in the world. When he was shipped back to New York, a crowd of four million awaited him.

Standing on the shoulders of the Wright Brothers, Lindbergh had just achieved the impossible. The groundbreaking solo flight established flight as the future of travel. The next year, Lindbergh would be *Time*'s first ever "Person of the Year." One year after his flight, commercial air travel increased 400 percent, with Pan Am (Pan American) being one of these early airlines.

Lindbergh Got Hearts Pumping

Ex-military pilots had little to do when the war was over. The army sold excess planes for $300, and aerial performances became a way to put food on the table for young pilots. Before his airmail days, Lindbergh was perhaps one of the most daring of these pilots. His adrenalin addiction led him to perform stunts such as killing the engine at low altitude, only to glide the rest of the way down, and even wing-walking (standing on the wings of an airborne plane). Lindbergh survived four plane crashes by parachuting out just in time.

Lindbergh's sister-in-law died of heart disease in 1930. This event inspired him to invent a way of keeping organs alive outside the body. Lindbergh joined forces with Nobel Prize-winning French surgeon Alexis Carrel and spent much of the early 1930s working on the project. By 1935, Lindbergh had developed a special pump that was capable of providing air and fluids to external organs, keeping them working while staying infection-free. The pump was hailed as a medical breakthrough, paving the way for the first true artificial organs. From organ experiments to air travel, Lindbergh had changed the medical field and aviation. While the air was being conquered, space, the next frontier, was about to see innovation.

Robert H. Goddard, the Father of the Space Age

Robert Hutchings Goddard is perhaps one of the most underrated dreamers of the early twentieth century. As a young boy in the 1880s, Goddard was obsessed with the skies and often studied the heavens with a telescope. Experiments would be conducted by the boy in-house.

Amusingly, this included an attempt to jump higher by holding a statically charged zinc battery. His mother, probably fed up with the scene, told him that he might fly away and never return if he succeeded. This caused Robert to promptly stop jumping.

After Goddard caused an explosion in the house while experimenting with chemicals, his father thought it might be a good idea to channel this passion in the direction of science.

At age sixteen, the young boy became interested in space after reading H. G. Wells' *The War of the Worlds* (the very same work whose live action radio broadcast would cause mass panic).

By 1914, he had come up with two patents for multi-stage rockets. In 1919, a compilation of Goddard's work was published. It was called *A Method of Reaching Extreme Altitudes* and described mathematical theories of rocket flight and even concepts of exploring Earth's atmosphere and space. The publication included a thought experiment on what it would take to send a rocket to the moon, backed up by solid mathematics. Goddard's early publication was hailed as one of the pioneering works of rocket science. However, this praise was only awarded in retrospect. At the time, sadly, most media outlets thought of his ideas as laughable. This caused the public to view Goddard as somewhat of a madman, even though his calculations were sound.

Due to the backlash, Goddard became a recluse and began to work mostly in secret. Another unfortunate reality was a lack of funding from third parties. Nobody wanted to give money to a lunatic. Despite the drawbacks, Goddard continued to work and, by 1926, he had developed the first liquid-fuel rocket, which took flight on March 16 of that year.

It only rose 12.5 meters (41 feet) and flew for 2.5 seconds, but it proved, for the first time, that liquid-fueled rockets were possible. After some tweaks, such as a controllable thrust nozzle, the rocket was then able to be guided.

After reading about one of Goddard's launches in 1929, Charles Lindbergh decided to step in and help. Lindbergh became a big fan of Goddard's work, managing to land him $100,000 ($1.8 million today) in funding. The advances in rocket technology that Goddard made, along with his impeccable vision, would later prove invaluable in the development of early missiles and space travel. When Apollo 8 became the first manned space mission to orbit the moon in 1968, Lindbergh sent the astronauts a message saying, "You have turned into reality the dream of Robert Goddard."

THE END OF AN ERA

"The Stock Market Only Goes Up!"

The mood of the early 1920s was one of unprecedented optimism. Commenting on America's economy, Calvin Coolidge (president 1923–1929) said, "The business of America is business." The public were on board with this idea, as life was improving. A common saying was "Keep cool with Coolidge."

Consumer credit was a brainchild of the 1920s. "Buy now, pay later. Live for the moment. Don't worry about the future," it advised. This mindset would bring forth the biggest financial disaster to date.

In search of ever more wealth, the average Joe was getting involved in the stock market for the first time. Telegraph-powered ticker tapes found their way into beauty parlors, ocean liners, nightclubs, and railroad stations, so everyone could keep up with the latest stock prices.

The culture of "buy now, pay later" eventually spread to the stock market, and people borrowed money to buy more stocks. By 1929, around 90 percent of the price of stocks consisted of borrowed money, and 40 percent of all money loaned from banks was for stocks. It didn't matter, though, because everyone was getting rich with money they didn't really have.

Party's Over

On October 23, 1929, for virtually the first time in a decade, prices on Wall Street began to drop. Motor companies fell first; no one knows why. As financial analyst Jim Rickards put it, figuring out financial collapses is like trying to figure out which snowflake caused the avalanche. We know the conditions surrounding why it happened—lots of snow, gravity, etc. But the exact snowflake that caused the chain reaction resulting in an avalanche? Impossible to know.

The fact that motor-company stock prices were falling wasn't such a big deal in itself, as market corrections are a natural occurrence, but President Hoover thought he should still address the nation, to prevent anxiety. On October 25, 1929, Hoover told the American public: "The fundamental business of the country is on a sound and prosperous basis." Four days later, it wasn't.

In the following days, a panic set in as people tried to sell...but no one would buy. Prices fell $2, $4, $10, with no signs of slowing. There were shrieks and gasps on the trading floor. Traders were terrified, and crowds began to gather on the streets outside the stock exchange buildings. On October 29, 1929, the bottom fell out and the stock market crashed.

It was the biggest financial disaster on record. All the investors who had borrowed money to buy stocks when they only seemed to go up, now had to pay that money back to the banks. The problem was, these investors didn't have any money.

All the machinery started to turn in reverse. The investors couldn't pay when the banks wanted their money back. The banks had to foot the bill and they started to go under, taking out much of the American public with them. Even trusted companies couldn't get loans to pay their workers, and layoffs, like an avalanche, swept across the country. In a flash, prosperity was gone.

These events affected both the US and Europe, as they were trading partners. Working-class people around the world were devastated. The mighty United States, the powerhouse of the post-war global economy, the consumer capital of the world, was groaning as it ground to a halt. The shudder of America's slowdown shook global trade, and soon the rest of the world began to feel it, as overseas companies shut their doors. Governments of the world could do nothing as, like dominoes, all industries began to falter.

An already struggling Germany was hit hard, and its people were looking for someone, *anyone*, to give them hope.

The decade which had started off with such promise and innovation was now ending very differently.

Sound, Vision, and Genius
1930-1939

By the 1930s, the collapse of many of the world's economies was in full swing. The Wall Street crash's follow-on effect killed trade. Many countries were economically destroyed, seemingly overnight.

In ports throughout the world, ships lay idle, as barely anyone needed to transport goods. Britain, the world's largest ship builder, was affected particularly badly. Like a disease, surrounding industries in the region became afflicted with a decline in business, and the knock-on effect of unemployment took hold of the entire economy.

The British government decided to cut spending and wait for nature to take its course. America did the same, and the result was horrific: bread lines stretched for blocks. The country was in dire straits, and the presence of slums increased. These slums were kindly labelled "Hoovervilles," after the sitting president. What had happened to the most prosperous nation on earth? How quickly it had fallen!

On the mere rumor of an open job, thousands of men would hitch rides on trains for miles, just for the opportunity to work. These included former scientists, doctors, and engineers; all were thrown into the same pit of desperation.

They say necessity is the mother of invention, and the hardest of times left us with one of the most creative inventions yet—the programmable computer. There will be more on that later in the chapter, but first, some examples of what other innovations were taking place.

The decade opened with the discovery of Pluto in 1930 by Clyde Tombaugh. In 1936, Kodachrome (invented by Eastman's company, Kodak) would bring color photography to the masses. In the same year, Russian-born Igor Sikorsky would invent the first practical helicopter featuring today's conventional design. It first took flight in 1939.

Hungarian brothers László and György Bíró invented the ballpoint pen, putting an end to smudge-prone and leaky fountain pens. Music performance would also get a boost, as both the first electric and bass guitars were invented within the decade. Both were attempts to make the instruments more audible in a live environment as electronic sound amplification began to improve. More important than all of these inventions of the '30s is the programmable computer.

THE FIRST COMPUTERS

When you think about it, a computer is a unique piece of technology. We're looking at an actual thinking machine that can perform calculations millions of times faster than any human could. It is, perhaps, the most important piece of technology of the twentieth century, so we'll spend some time here.

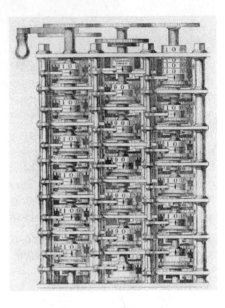

To understand how the computer came about, we have to take a look at the bigger picture—a timeline that stretches from the seeds of the idea to the first working model.

Charles Babbage and the Automatic Calculating Machine

As we saw in chapter 4, a "computer" was an actual person. The word "computer" originated in the 1640s, and it means "one who calculates." In the seventeenth century, it was a person whose profession was computing calculations using tables. Although boring, it was truly an important job, as mathematical tables weren't the irrelevant chore we view them as today. Back then, tables were the powerhouse of science, engineering, finance, and even economics, and the only way a table could be made was by a team of talented mathematicians. Occasionally, though, even they made mistakes.

In the early 1820s, a keen mathematician and philosopher named Charles Babbage was on the receiving end of human error.

Charles was getting annoyed because he repeatedly found mistakes in log tables used for astronomy. As annoying as that was, it wasn't the end of the world, but other mistakes had real-life consequences.

For example, ships would sometimes sail off course and run aground because of mistakes in their navigation tables. This was chaos, and Babbage knew a fix had to be found. The solutions that Babbage conceived would, for the first time, bring forth the key ideas of modern computing.

While thinking about the calculation problem, Babbage was about to have the mother of all ideas.

He asked: If physical machines can do physical work...why can't physical machines do *mental* work?

With this question, the idea of a machine that could calculate all the results of a table and print them out correctly was born. He called this invention a "difference engine."

The function of the difference engine was to calculate polynomial operations (logarithms, sines, cosines, and tangents) by only using addition and subtraction.

The machine was completely mechanical in nature, featuring vertical shafts with disks that could display any number. The disks were attached to cogs that mechanically performed addition or subtraction on the shafts next to them.

By 1822, Babbage had built part of a working model capable of calculating polynomial equations, error-free and steam-powered. This was unbelievable at the time, a physical machine that could do complex mental work—a dream come true!

Babbage would demonstrate his machine at parties in the sight of highbrow guests such as lords, poets, statesmen, and industrialists. At one such party, a young lady by the name of Ada Byron (later known as Ada Lovelace) witnessed the machine, and she absolutely loved it. In collaboration with Babbage, Ada made her own major contribution to history, and there will be more on her shortly.

Unfortunately for Babbage, the tools of the day proved unfit to finish a larger-scale machine. Determined, in 1834 he set out to design a general-purpose version that could calculate *anything*, not just polynomials. This was to be called an "analytical engine."

The concept—known as a general-purpose computer—was one hundred years ahead of its time.

The Analytical Engine: Early Plans for a Computer

Completely mechanical in nature, the plans for this analytical engine included a "mill," which controlled, rotated, and positioned all the mechanical numbers on the wheels.

Essentially, this "mill" is the equivalent of a "central processing unit" (or CPU) in a modern computer (we'll revisit the CPU later in chapter 9).

The design also included a "store" where all the correct numbers landed (what we would call "computer memory" today).

The analytical engine was to be fed with punch-card instructions to tell it what to do, and even featured a printer as well as a curve plotter. The analytical engine, with a "CPU" and "memory" that could be programmed, represented the first complete plans for a computer. However, it was never built, as it was far too complicated to be constructed.

Ada Lovelace: True New Thinking

Ada Lovelace achieved an important milestone in history despite the greatest odds. She was the first to realize that computers had applications beyond bland calculations. She also published the first ever algorithm (though hypothetical) and is considered the very first computer programmer.

The approach she used in her work was definitely new thinking.

Lovelace's life reads like the script of a melancholy movie. She was born in 1815 in London, which is fine, but it started to go downhill from that point on. Her father's first words to her were, "Oh! What an implement of torture have I acquired in you!"

Unsurprisingly, her parents separated when she was one month old, and Lovelace would never see her father again. He died when she was eight.

From age four, the young girl was privately schooled by some of the best tutors in the fields of mathematics and science. Lovelace's mother thought teaching her math from an early age would drive out any insanity inherited from her father. It must have been a bitter divorce, to say the least!

At seventeen, Lovelace began to show great promise in math, and was hailed as an original thinker. To her, imagination and intuition were a must when it came to mathematics. She was just eighteen when, in 1833, Babbage showed her a prototype of his "difference engine." Lovelace was entranced. They would begin working side by side, and Lovelace's mind would impress Babbage, who gave her the nickname "the Enchantress of Numbers."

Lovelace was well-liked and often rubbed shoulders with the likes of Charles Wheatstone (the telegraph) and Michael Faraday (the DC electric motor), both geniuses in their own right, as we saw in chapter 2. Imagine all the intellectual brilliance in that room! Lovelace was particularly interesting, not only because of her different ideas in math, but also because of how she saw computing. While others, such as Babbage, saw a computer as a cold calculating machine, she envisioned a computer to be much more than that.

Ada Lovelace had the foresight to imagine it as a collaborative tool, providing a way that text, images, and sound could be converted into a digital form and manipulated by a mechanical computer.

In 1843, Lovelace was given the task of translating an article on Babbage's analytical engine from Italian to English. Interestingly, the article was written by Luigi Menabrea, an Italian military engineer and future prime minister.

While translating the paper, she decided to make some notes. As she began to write, the notes soon took on a life of their own, growing three times longer than the article itself. The result was something truly special, something that would become a huge moment in the history of technology. Lovelace's notes included a method of instructions "programming" the analytical engine to calculate a mathematical problem. This "program" would have worked if a full-scale analytical engine had ever been built. Lovelace had just written the first computer program algorithm.

Lovelace's thinking was so far ahead of her time that it would take almost one hundred years for her ideas to be recognized. It wasn't until Lovelace's notes regarding Babbage's analytical engine were published, in 1953, that the world realized that the first computer program had been written well before anyone could have imagined.

Despite this, to this day, Lovelace is not well known to the wider public. Sadly, Lovelace died from cancer at age thirty-six. She was the same age as the father she never knew. She requested to be buried next to him.

Don't Tell My Husband about My Gambling Problem

Lovelace's love for numbers extended into the world of gambling. She attempted to come up with a mathematical model to predict winning horses and win vast sums of money. It didn't work. In fact, it did the opposite.

She ended up losing more than £3,000 (£311,000/$431,000 today) on the horses during the later 1840s. This vast amount of debt left her no choice but to admit the blunder to her husband, who was probably less than impressed.

Alan Turing: The Modern Computer Visualized

Born on June 23, 1912 in London, Alan Turing was a brilliant, inquisitive, but somber individual. Later in life, he was a pioneer of computer science and artificial intelligence. In middle school, Turing could solve calculus problems in his head, before he had even been taught calculus. By the time he reached college, he had a keen interest in quantum physics—particularly the mathematics behind it.

Calculus was as hard as eating cake for Turing. While on a jog in the summer of 1935, a twenty-three-year-old Turing envisioned the concept of a general-purpose computer (independently of Babbage)... in his head.

It was a mental idea of a machine that could solve problems with only a finite number of controls and instructions. He called this machine the Logical Computing Machine (later Turing machine).

But what problem is being solved here? This is where things get interesting. The problem that's being solved is actually a philosophical one: "Can mathematical statements be proven true or false by a set of rules?"

The technical name for the idea of being able to solve all mathematical problems only using a defined set of rules is called the "decidability" problem. It was first asked by German Mathematician David Hilbert in 1928. It basically asks:

"Could there exist, at least in principle, a definite method or procedure by which all mathematical questions could be decided?"

Basically, is there a method to solve *any* math problem using a finite number of steps? This was the same problem that Babbage had sought to solve mechanically using the analytical engine, except Turing was examining this from a theoretical perspective.

It took Turing to come up with a more complete solution than Babbage's, in the form of a thought experiment: the Turing machine. This machine, in theory, would compute *any* mathematical problem, using only a set of rules. In other words, it was the concept of a general-purpose computer, but more defined this time.

The Turing machine provided the blueprint for the way modern digital computers work. Even today, we use this thought experiment to measure the strength of computer programs. When a program is capable of doing what a Turing machine can do, it's called "Turing-complete" and is at the highest level of programming-language strength. There's never been a way to do *more* than what a Turing machine can do.

Later on, Turing aided the Allied victory in World War II by designing a code-breaking machine called the Bombe. This machine allowed Great Britain to know when the Germans were planning an attack, giving them a critical advantage in the war.

Turing would go on to develop other ideas, such as the Turing test for artificial intelligence. In 1950, Turing's paper *Computing Machinery and Intelligence* described a test that hinted that computers could possess intelligence. A test participant would talk to a person via a keyboard, and at some point, unknown

Z1 Computer

to the participant, the person on the other end would be replaced by a computer.

If the participant couldn't detect the change (that is, the computer's responses were indistinguishable from a human's), that computer passed the Turing test. The Turing test was only definitively passed in 2018 by Google, but only in the narrow scope of phone conversation and in the context of making booking reservations.

In a running theme with all the great pioneers of computing, Alan Turing would meet a sad end. In 1952, his home was broken into. A police investigation into the burglary revealed evidence that Turing was engaged in a homosexual relationship, an offence punishable by imprisonment.

After the revelation and his arrest, Turing's reputation suffered. Overcome by the circumstances, he committed suicide using cyanide. Alan was just forty-two.

Konrad Zuse: First Programmable Computer

All the ideas from the previous stories lead up to the first *physical* construction of a freely programmable computer. Its inventor was a man who was too lazy to do calculations himself.

In 1935, Konrad Zuse was a German engineer at the Henschel Aircraft Company. One of his jobs was to analyze the stresses in aircraft. Solving a stress equation problem with six variables took over a day to do by hand; twenty-five variables would take over a year. This doesn't even mention time lost to mistakes.

Zuse wanted to make his life easier by building a machine to do the calculations for him. "I was too lazy to do the work," Konrad said in a 1992 interview, "so I invented the computer."

As a kid, Zuse loved Meccano sets, as well as the artistic and technical sides of structures. He chose civil engineering for the combination of these aspects.

When he graduated, Zuse found himself in an unpleasant situation. In order to calculate the forces and stresses on buildings and bridges, he had to go through hellish boredom equivalent to a thousand episodes of the *Big Bang Theory*. The repetitive calculations were filled out in what seemed like endless rows.

Zuse's interest in Meccano sets as a child must have come back to him, because he got the bright idea of using a giant set to automate his work. Zuse believed in his idea so much that in 1936, he quit his job to move back home with his parents to work on his machine. I'd imagine his parents were quietly disappointed.

Hitler received with great fanfare

After a while, Zuse invited some friends from university to help out. They all thought the project was pretty cool, but not even Zuse had an idea of how the mechanical calculators at the time worked. Thanks to his ignorance, he managed to see the problem from a totally new perspective and started building an automatic calculator from scratch.

He reinvented the wheel but did it differently. You see, at the time, mechanical calculators calculated in decimals. Zuse bypassed this and chose to use the binary system (the machine's switches would just go to position 1 or position 0, instead of using complex gearing to switch from 0 through 9). This proved to be much simpler and enabled more complicated calculations.

The hardest part for Zuse wasn't figuring out the logic and mathematics of how to put the thing together: it was the task of building it. The Z1 (the name he'd given to his eventual machine) consisted of one thousand thin, slotted metal plates that made up its memory. The program was put into the machine by a tape reader that read old 35-mm film stock with holes punched in it. It also had a keyboard and displayed its results via a row of lights. The computer could store sixty-four numbers and could perform addition, subtraction, multiplication, division, and square-root operations.

In 1938, with the help of his friends, Zuse somehow got the Z1 up and running. He tried calculating a 3x3 matrix and it worked. The very first freely programmable computer in history had just been turned on! Unfortunately, it broke immediately. This turned out to be a constant: the Z1 was too unreliable and prone to breaking down. Zuse worked on his idea until he had a better design in the form of the Z3, which used telephone relay switches for the 1 and 0 positions.

A machine that could calculate mathematical problems would prove invaluable in war. And war was about to come.

RISING WAR TENSIONS

In 1934, Germany elected Adolf Hitler as chancellor. In part, this was a response to the terrible economic conditions in Germany at the time. He gained popularity by first attacking the Treaty of Versailles (and the punishment it forced upon Germany).

The people wanted change, and the German Socialist Workers Party (Hitler's party, the Nazi party) seemed to be getting things done. In just two years, seven million Germans were put back to work. Manufacturing was booming, new civil projects such as the Autobahn were being built, and most importantly for many, Germans could have a sense of national pride again. Even for the children of the day, marching as one of the so-called "Hitler Youth," under a swastika flag, was an honor and a joy. The symbol they were marching under didn't mean much at the time. It was simply the symbol of a nation that was on its way up and would no longer take things lying down. Unfortunately, the people's pride would be turned into something more sinister when propaganda outlets convinced a large proportion of the German population that they were superior to all. Soon, the focus was shifted to the "problematic" Jews, Poles, and Slavs, as well as the disabled, living in Germany.

Volkswagen: From Hitler to Hippies

Hitler had many plans for the people of Germany. To communicate his message to the masses, affordable radio sets called the Volksempfänger (German for "people's receiver") were to be in every home. Hitler had learned to use the emerging technology of radio to distribute propaganda, despite the radios' not-so-catchy name.

Apart from radios, he would also give the people something else: a car.

In the 1930s, Germany was far behind in the adoption of motor transport, compared to America. German cars were only for the rich, and a mere one in fifty Germans owned one. In America, this number was around one in four.

As part of a new vision for Germany, Hitler wanted a car that was affordable for every German. It was to be called the "people's car," or Volkswagen. This idea wasn't unique to Nazism, of course. The American Ford Model T, which came to prominence in the last chapter, showed us as much.

In 1938, the first Volkswagen factory was built. Hitler laid the cornerstone and named this particular Volkswagen model during a speech. The car that was to become the friendly Beetle started out its life with a slightly stronger name. It was named *Kraft-durch-Freude-Wagen* ("Strength Through Joy Car").

The Strength Through Joy Car (is the name growing on you yet?) was to start production in 1939. However, only 210 of these cars were produced before the outbreak of World War II halted production. In the end, the cars were only given to high-ranking members of the military, while the factory proceeded to make four-wheel-drive military versions. The German civilians initially missed out.

After the war, as part of the conditions for ending the conflict, the Volkswagen factory was given to the Americans. Henry Ford was offered the war-damaged factory (with an undetonated bomb stuck in the roof) for free, but he wasn't interested. The factory was then handed over to the British. The British Automotive manufacturers looked at the little Beetle, and thought it was one of the worst things they'd ever seen.

They stated: "The vehicle does not meet the fundamental technical requirement of a motor-car…it is quite unattractive to the average buyer… To build the car commercially would be a completely uneconomic enterprise."

A British Army officer by the name of Ivan Hirst was set to the task of cleaning up the factory and putting it back into working order. He was so successful that he was made its director in 1949, and by 1955, 1 million Beetle cars had been produced. Despite its being "unattractive to the average buyer," the car did well, and was superior in performance and reliability compared to its European competition.

International exports grew in Europe, but in the United States, the cars were a laughingstock compared to the powerful, large, flare-finned monsters on wheels available there. If Volkswagen could convince the Americans to buy the Beetle, it would open the company up to the largest consumer market on the planet. But how were they to do that?

The company knew that the Americans would never believe this ugly little car was more powerful than their cars, so why not *own* the ugliness? The idea was brilliant. In 1959, it was put into practice. Some advertising slogans of the day included "Ugly is only skin deep" and "A face only a mother could love." It worked, and soon, the car had a cult following in the United States. The VW Beetle, and later the Kombi van, became an icon of the counterculture during the 1960s. The company had literally gone from Hitler to Hippies.

The Beetle would go on to become one of the longest-running car nameplates in history. Over twenty-one million units of the original model were produced, with relatively few changes, from 1938–2003. The original model was replaced by the "New Beetle" in 1998. The Volkswagen group is now the largest automaker in terms of sales, owning Audi, Lamborghini, Bentley, Bugatti, Ducati, Porsche, Skoda, and SEAT.

The Beetle, which started off as a harmless little car for the German people, would have been a fine end for Hitler; but unfortunately, Hitler didn't stop there.

War Is the Mother of Invention
1940–1949

As we enter the 1940s, the world moves into a critical period. Despite this, life goes on as normal for some time. Swing is the most popular genre of music, and Walt Disney is on top of his game with movies like *Dumbo*, *Bambi*, and *Fantasia*—all technically groundbreaking films, produced within a couple of years.

The decade was indeed full of history-altering moments, but it was also one of dichotomies: In the second half of the 1940s, there were times of great prosperity and optimism in the United States, while a dark mood of somber reflection and rebuilding enveloped Europe and Japan. We were given light-hearted inventions like the Frisbee, the Slinky, and M&M's, but also the atomic bomb and LSD.

The end of the war in 1945 saw the biggest spike in population ever, as young people saw a brighter future. These babies of optimism would be known as the Baby Boomers. Those born in 1946 include familiar names like Tommy Lee Jones, George W. Bush, Freddie Mercury, Steven Spielberg, Sylvester Stallone, and Donald Trump.

WORLD WAR II

The First World War was said to be the war to end all wars. Sadly, that was not the case. In 1939, the world would go to war once more—only, this time, soldiers would be armed with better weapons, faster planes, terrifying tanks, and the destructive atomic bomb.

The war would drag out over five long years and involve over 100 million people from thirty countries, and when it was all over, 50 to 85 million human lives would be numbered as casualties.

One war grew from the last, as German unrest led to Adolf Hitler's rise to power and subsequent attempt to impose his will on the rest of Europe. As alluded to earlier, Hitler's Nazi party was built on the ideas of Aryan superiority and made scapegoats of several minorities. The main focus was on Germany's Jewish population, who would be exterminated in horrifying numbers over the course of the war.

Germany had steadily been increasing its influence. Having created an axis of power with Italy and Japan, Germany invaded Austria and Czechoslovakia, who then also joined with the Germans. On the September 1, 1939, despite cries of warning from Britain and France, Germany invaded Poland. This was the official start of World War II.

The invasion of Poland was quick: The Germans used a tactic they would refer to as a Blitzkrieg, which involved lightning-fast attacks with Germany's technologically superior tanks and planes. A sea of German tanks rode into the Polish country side, joined by German dive bombers from the air. The Polish army, who were still using horses and lances, didn't stand a chance. It was clear that this was going to be a new type of war.

After Poland, Germany quickly amassed land, running riot through Denmark, Norway, Belgium, and the Netherlands before finally taking France.

Germany then set its sights on Britain, but the country's island geography, as well as its superiority both at sea and in the air, held off the German attacks, though not without severe destruction caused by German bombing.

Meanwhile, on the Eastern front, Hitler's shaky truce with Joseph Stalin and the Soviet Union was starting to crumble, and the two would butt heads. The early stages of the German invasion of Russia quickly took ground. Russia took severe losses, including 20 million citizens over the course of the war. But in the end, the Russian winter would prove too much for the German soldiers, and all progress would cease just outside of Moscow.

On December 7, 1941, a fleet of Imperial bombers were flying toward an American naval base in Pearl Harbor, Hawaii. In a bright flash of light, while an American naval band was still playing, the harbor was attacked by the Japanese, who had joined the side of the Germans and Italians. The bombing was an attempt to disable the US Navy, who presented an obstacle to the Japanese taking nearby regions such as the Philippines and Vietnam. The Pearl Harbor attack would officially pull America into the war.

Meanwhile, in New Mexico, some of the best scientists and engineers in the world were working on a well-guarded secret project in the desert, dubbed the "Manhattan Project." It was based on the theory that splitting an atom could release huge amounts of energy: a bomb powered by atoms. Albert Einstein had previously described this in a 1939 letter to President Roosevelt. The Manhattan Project team, led by Robert Oppenheimer, were now in a race against Germany to create the most powerful weapon the world had ever seen.

While these scientists were working to create the first atomic bomb, the Allied forces (UK, France, US, Australia, Canada, etc.) managed to take Italian-occupied regions of North Africa and gain ground in Italy itself. On June 6, 1944, the Allies launched the D-Day invasion of Normandy (France) and would soon liberate Paris. The Allies moved into Germany from the west, east, and south, liberating nations along the way and freeing survivors from Nazi concentration camps. On May 8, 1945, the Germans surrendered.

On August 6, 1945, the Americans dropped the atomic bomb on Hiroshima, Japan, and, days later, this was followed by a second attack on Nagasaki. Those caught in the blast were vaporized—all that was left were eerie shadows and dust. Between the blast and the ensuing radiation, over 200,000 people lost their lives. Japan finally surrendered.

The war was over, but at a terrible cost. Germany was split in two: East Germany was run by the Soviet Union and West Germany was held by the Allies, an arrangement that would last until 1990. The conflict was the most destructive period humanity had ever faced. It is difficult to see any kind of silver lining in all of this loss and bloodshed but—perhaps out of sheer desperation to see the nightmare end— the war did push technology to new heights. Humanity rose from the ashes with advances in computer technology, rocketry, radar, aircraft, cryptography, and the transistor. Let's see how.

Why Did My Snack Melt? The Microwave Oven (1945)

Next time you're about to heat up some leftovers in your microwave, think about this story:

Radar (or Radio Detection and Ranging) was used during the war. Like a searchlight, a short pulse of microwaves (sub-optical electromagnetic frequencies) would go out to an object and be reflected back to the receiver, allowing it to "see" the object and gauge its distance and position.

Originally invented by the British, radar was perfected in the US for use against the Nazis. This allowed Allied forces to "see" German planes long before the German planes could see the Allies. It's said that the quick development of radar technology during the war tipped the scales in favor of the Allies. In the United States, a military manufacturing company called Raytheon was the main producer of critical radar equipment (called a magnetron) for the war effort.

In early 1945, Percy Spencer, a Raytheon engineer, was working on improving the company's radar equipment. It had been a long day's work, and Spencer's stomach was beginning to grumble; it was time for a snack. When Spencer reached into his pocket to pull out his tasty treat, he was surprised to find that what had once been a chocolate bar was now a completely melted mess. Spencer quickly realized the microwaves from the radar's magnetrons must have heated it up. He knew he had stumbled across something big. Excited, Spencer asked a research assistant to bring him a bag of corn kernels. When the kernels were put on a table near one of the magnetrons, they began to pop. It was confirmed: the microwaves used in radar equipment could cook food!

It turned out the microwaves gave energy to the molecules in food, causing them to vibrate rapidly. Vibrating molecules are essentially what heat is: when something warms up, all that's happening is that the molecules making up the material are vibrating at a faster rate. Hence, microwaves could heat food in a totally new way. The American public would love an invention that could heat up food in a fraction of the usual time.

As the war ended, Laurence Marshall, then CEO of Raytheon, thought the microwave idea was great. He insisted that Spencer put all his efforts into creating a usable product for the public. Spencer was given a team of engineers and set to work.

After some time, the team came up with a product, but it wasn't exactly portable. It was the size of a refrigerator: 1.8 meters (6 feet) tall and weighing in at 270 kg (600 pounds). This beast put out 3,000 watts of power and was dubbed the Radarange (later to be known as "the microwave"). In the information booklet, there were some cooking suggestions that seem a little odd today: "Well done steak in fifty seconds? No problem. Fried eggs? Done before you can count to twelve."

In 1947, Raytheon released a commercial version that was slightly more compact, for $3,000 ($33,000 today). As you can imagine, this was too expensive for the average person. For this reason, the device was sold to restaurants and other corporate food venues. Over the years, it was noted that an expensive military-grade magnetron was overkill when it came to heating up some leftover chicken. Gradually, the microwave was reduced in heft, and by the late 1960s had shrunk down to the form we are familiar with today.

The war had an impact much greater than heating up leftovers, however—it also spawned a giant leap in aircraft technology.

THE JET AGE

After Germany was defeated in World War I, the Treaty of Versailles ordered the country to cease all development of aircraft technologies. However, as they were not yet known at the time of the treaty, jet and rocket technologies were never mentioned in this agreement.

At the German university of Göttingen, a young engineering student by the name of Hans von Ohain was looking into the idea of a jet engine. At twenty-one, he was already a brilliant physicist. With the encouragement of his professor, von Ohain started work on building a prototype of a new propulsion system.

Meanwhile, in Britain, a twenty-six-year-old air force pilot, Frank Whittle, was investigating the same problem of jet propulsion with no clue about what was taking place in Germany. In 1932, he filed a patent for the world's first jet engine. Despite Whittle's revolutionary thinking, the British military didn't have faith that his idea would work. As a result, Whittle was starved of funding.

If the British had embraced his jet engine, they could have had a deterrent to stop Hitler from marching into Poland in 1939 and risking a war with Britain and France. To add insult to injury, the patents from Whittle were published in technical documents around the world, including in Germany. Some historians theorize that the young university student, von Ohain, may have read these patents and gotten a few ideas to incorporate into his own designs.

Either way, by 1935, von Ohain had a working prototype jet engine. He and his professor went to the

German aerospace company, Heinkel, who immediately saw the potential and asked for a full-scale engine.

Back in Britain, Whittle had founded his own small company and tested his jet in 1937. Early testing was full of setbacks and failures. One case included Whittle losing control of the thrust function of the test engine, causing it to continuously rev higher. At that point, most of the staff fled the room, but Whittle stayed put, interested to see what would happen. The engine eventually exploded and, fortunately, Whittle was uninjured.

The British government weren't too pleased with such spectacles and were even less convinced by this dangerous new jet-engine idea.

With Hitler pushing for a one-seater jet aircraft to make the Nazi air force a formidable threat, the Heinkel firm

ENIAC Computer

pushed on urgently with the project. In 1939, just one month before the outbreak of the war, a prototype jet aircraft—the He 178—went airborne.

The Germans had beaten the British in the secret race to have a jet in the sky. At the outbreak of the war, however, the Nazis held authority over both land and sky using just traditional propeller-driven aircraft. Surprisingly, this was a blessing in disguise, as the Germans soon lost interest in the new propulsion-jet aircraft—their propeller planes were doing just fine, after all.

Still, just because the military lost interest in jets didn't mean everyone in Germany did. A forty-one-year old aircraft designer, Wilhelm Messerschmitt, set his sights on the prize of building a jet *fighter*. The result was the Messerschmitt 262. It could fly at 965 km (600 miles) per hour and made it into mass production. By 1944, it was in operation. It was the most formidable aircraft of the war. It flew 160 km (100 miles) per hour faster than the Allied planes escorting and protecting their bombers. Because the technology was so new, however, the engines only had a lifespan of twenty-five hours before destroying themselves.

The Germans also experimented with some slightly odd ideas. One was the Messerschmitt 163 rocket-powered plane. It had swept wings, no tail, and no space for wheels after take-off, so they dropped off and fell to the ground during flight. The fuel in the 163 was so volatile it could dissolve the pilots if they weren't wearing protective gear. To make matters worse, there were no two-seater training versions, so the pilot's terrifying first flight had to be made alone.

The 163s could climb around four and a half times faster than any traditional aircraft. By the end of the war, they were used to surprise Allied bombers. They could outmaneuver and outrun any other aircraft but could only stay in the air for three minutes.

The planes were not fun to fly. Since its wheels were discarded once airborne, 80 percent of losses happened during takeoff and landing.

After dragging their feet, the British eventually came out with their own jet, the Gloster Meteor, in 1944. It was slower than the German jet, but Whittle's engine (which was used in the Meteor) was more powerful and more reliable than the Germans'. In the end, it barely mattered for either side, because the war ended within a year. Frank Whittle did later receive a knighthood from the Queen for his contributions, however.

THE BIRTH OF MODERN COMPUTING

From 1940–1945 the United States and Britain worked independently on creating computers to gain an advantage during the war. By 1951, there was a commercial computer on the market. Information could now be processed and manipulated automatically. Here's the story of how it all happened.

First All-Electronic Digital Computer

John Vincent Atanasoff was a professor at Iowa State University. He noticed that his students would take weeks to manually calculate the linear equations used in their physics courses. Like Konrad Zuse in the last chapter (who also used linear equations in his stress calculations), Atanasoff grew impatient with this process. In 1934, the Iowa professor began to explore the idea of automatic computing.

In 1939, Atanasoff would enlist the help of a graduate student by the name of Clifford Berry.

By 1942, the pair had come up with the Atanasoff Berry Computer (ABC). It was a breakthrough because it used electronic vacuum tubes to perform calculations, rather than wheels or mechanical switches (like Konrad Zuse's computer). However, the ABC wasn't programmable or general-purpose in nature.

How Computers Ran on Basically Light Bulbs

The vacuum tube, an emerging technology that had been improved in war, was looking like it could find a new home.

A vacuum tube is a vacuum space enclosed in glass, similar to a light bulb, but with electrical connection points at the top and bottom. When the tube is heated, electrons can flow in one direction, from one side of the tube to the other. This function can be manipulated to act as an "on/off" switch that can represent 0s and 1s (we call these bits). These bits can be used to run code or organized into clusters of bytes (8 bits to a byte) to store data. Vacuum tubes were faster than mechanical switches, which made them a prime candidate for computers at the time.

Vacuum tubes were also used in radios, televisions, and early radar equipment before being replaced by the transistor, which we'll get to in the next chapter.

The war interrupted Berry and Atanasoff's progress, as Atanasoff went to work for the US Navy. After the war, the ABC computer was salvaged for parts and was never patented. The ABC largely went unnoticed until 1973, when a court case over a patent dispute with the ENIAC (a later US military computer) recognized the work of Atanasoff and Berry. In the ruling, it was stated that the basic idea for modern computing came from Dr. John Vincent Atanasoff. Still, many people doubted that his original machine worked, so in 2010 a group of staff and students at Iowa State set out to rebuild the ABC.

They used only the original plans and genuine parts from the 1940s (including old vacuum tubes). The machine did in fact work, proving that the origin of digital computing was the ABC. Although revolutionary, the ABC was not a general-purpose machine (freely programmable). It was designed only to solve linear equations. For a computer that could calculate technically anything, we must take a look at the ENIAC.

ENIAC (The First Electronic General-Purpose Computer)

Though Turing and Babbage were the first to envision the general-purpose (calculate anything) computer in theory, such a machine was never built—at least electronically. The new thinkers who would materialize these visions were just around the corner.

Firing Tables

We've learned how much of a pain tables can be from the story of Charles Babbage. However, tables were still being used by US military in World War II, so they could aim and fire correctly depending on wind and temperature conditions. Creating firing tables required great amounts of effort and used time-consuming, mechanical adding machines.

By the middle of the war, there weren't enough tables to meet demand. Without firing tables, ground gunners would be firing their artillery shells blindly, which was costly and a waste of artillery.

Enter physicist John Mauchly from the University of Pennsylvania. The physicist had a radical idea—a giant electronic computer that could figure out an artillery trajectory in 100 seconds, instead of weeks. It was essentially the original supercomputer. Taking just one hundred seconds to do a week's worth of work sounded like insanity. But, desperate to solve the problem, in 1943, the US military reluctantly coughed up $500,000 ($8.8 million) for the now top-secret project.

Why Wouldn't the Damn Thing Just Work?

Mauchly and graduate student J. Presper Eckert set to work on this Electronic Numeric Integrator and Computer (ENIAC). Driven by news of their fellow men dying in battle, the pair toiled tirelessly to complete the project.

It was an incredible task. They had to design, build, test, and learn how to program the machine from scratch as quickly as possible. By the end of their labor, the diligent pair had a monster 15-meter-long (50-foot) machine that weighed 30 tons. The worst part? Doing all the switching (shifting between "1s" and "0s" to perform calculations) were 17,500 delicate, prone-to-breaking-*JUST*-when-you-don't-want-them-to vacuum tubes. We touched on vacuum tubes earlier: although an advance from their mechanical predecessors, if any one of them failed (burned out like a light bulb from overheating), the whole machine wouldn't work.

This happened so often that even with precautions, the thing would only function for about ten to thirty minutes before breaking down. Regardless, it could perform 5,000 additions, 357 multiplications, and 38 divisions per second. This was the most advanced computer ever made at the time, despite its many shortcomings.

Apart from the burning-out problem, ENIAC also had practically no memory and had to be laboriously rewired (unplugging, rearranging, and re-plugging hundreds of cables) for each new problem it was asked to solve—interestingly, thanks to computer memory, this process would be as simple as double-clicking today.

Even so, ENIAC was the very first physical computer (as opposed to visualized computer) that could calculate anything it was given. The only problem? ENIAC was completed two months *after* the Japanese surrendered and the war was over.

UNIVAC (First Commercial General-Purpose Computer)

After the war and the ENIAC project, Mauchly and Eckert decided to use this new computer technology to make life easier. They wanted to build a computer for the world of business.

In March of 1951, after six years of work, the university pair sold the first general-purpose computer to the United States Census Bureau: UNIVAC. Unlike the earlier rewiring mess that was ENIAC, UNIVAC could be programmed for many different tasks much more easily. Tape drives held the data and results were printed quickly. Impressive for 1951, but hardly anyone took notice.

This muted reaction would literally change overnight as the computer burst into the spotlight. In 1952, CBS and Remington Rand wanted to do the ultimate public-relations stunt: a computer was to predict the outcome of the presidential election between Dwight Eisenhower and Adlai Stevenson. This would be a first.

As the votes came in, UNIVAC predicted a landslide victory for Eisenhower, even though poll numbers said the result was too close to call. When CBS heard what the computer had printed out, they insisted that the UNIVAC team check their machine because it was *definitely* malfunctioning. But sure enough, Eisenhower won by a landslide. UNIVAC had predicted the victory to within 1 percent of the real numbers. The computer had arrived. As amazing as this was, these machines were still powered by temperamental vacuum tubes. It would take an invention called the transistor to change that.

THE TRANSISTOR: PROMISE AND BETRAYAL

Computers may have entered the picture, but the problematic vacuum tubes that made them were power-hungry, unreliable, and expensive. Enter the transistor.

The transistor may be the most important invention of the twentieth century. The original purpose of the device was to simply amplify (strengthen) electrical signals, but it soon turned into something much greater. No one could have expected the revolution something so physically small would bring to business, culture. and society.

The transistor is an integral part of any digital device in existence today. Our computers have billions of them tucked away in their chips. In addition to amplifying signals, a transistor can *also* be a switch that controls the flow of electrons, just like a vacuum tube. It can either be in an "on" ("1") state or an "off" ("0") state.

These on/off states can be translated into binary code, which is how all our data currently exists. By switching between states billions of times a second, these little transistors allow data to be processed and manipulated, giving us the digital world of today. They're in everything with a chip in it: from TVs, radios, and VR headsets to computers, cars, and mobile phones. Today's transistors can be as small as a few nanometers, but the first devices started off about as large as the palm of your hand.

The transistor's story is one you wouldn't expect: a story of promise, jealousy and conflict between three men.

THE TRANSISTOR'S ORIGINS

The story of the transistor actually starts earlier than the drama between the trio, all the way back in 1907. At that time, Alexander

Graham Bell's company AT&T had a problem on their hands. Bell's patents on the telephone were about to expire and the company was facing a great deal of competition. The company even brought its former president, Theodore Vail, out of retirement to help. As the patents would be lost anyway, Vail decided the company must think of something new. It must offer something no other company their size could dream of offering—a transcontinental telephone service *across* the United States! It was an outlandish boast that hadn't yet been proven possible.

However, just one year earlier, American inventor Lee de Forest had come up with a modification to the vacuum tube that could amplify electrical signals. Since audio electrical signals could pass through on telephone lines, it could amplify voice. This was just what AT&T had been looking for. Their dream was not so far away after all…

If amplifiers could be installed all along the length of phone lines, the signal could go right across the country, loud and clear. AT&T jumped at the opportunity and bought de Forest's patent, making their own drastic improvements. Their changes allowed the signal to be amplified regularly along the line, meaning that a telephone conversation could now travel across any distance.

Over time, though, a major problem developed. It was those vacuum tubes again. As we know, the tubes (that made the amplification possible) use too much power and produce too much heat, making them extremely unreliable.

In the 1930s, Bell Labs' director of research (Bell Labs was the research branch of AT&T), Mervin Kelly, recognized that a better device was needed for the telephone business to continue to grow. He had a feeling that the solution might lie in a peculiar class of materials called semiconductors.

Semiconductors have an electrical conductivity that lies between that of a conductor (like copper and silver) and an insulator (like glass or plastic). During the war, the rapid advancement of radar technology spurred research in new ways to achieve faster response times. Vacuum tubes weren't up to the task, so silicon materials were examined. As it turns out, silicon (the reason for Silicon Valley's name) is a great semiconductor and worked flawlessly in radar systems. What wasn't certain was whether semiconductors could work in other systems.

In 1945, after the war, Mervin Kelly decided to put a dream team of specialists together to find a solution to the vacuum tube problem, via semiconductors. This special material led to the creation of the inimitable transistor.

The characters Kelly selected for the task couldn't have been more different.

The Semiconductor Transistor Dream Team

First there was Walter Bratton, the oldest of the three and an experimental physicist who was great with his hands and could build anything. Next, John Bardeen, one of the twentieth century's greatest theoretical physicists, who skipped three grades and entered college at age fifteen. Despite his intelligence, Bardeen was mild-mannered and meek. Last was Bill Shockley, the youngest of the three and a team leader. He was a brilliant specialist in science theory, while also driven and fiercely competitive. This last characteristic would prove to be his undoing.

Just before the research team was assembled, Shockley built a crude model of something he thought might work like (what was later termed) a transistor and began to test it. For many years, he had envisioned becoming famous by perfecting this transistor system using semiconductors. Despite his ideas seeming solid on paper, the device didn't work.

Shockley took these ideas with him to the research team and re-examined them there. Thinking that he'd done some of the math wrong, he asked his colleague John Bardeen to take another look as his work. It was correct... Why didn't it work? Shockley asked Bardeen and Bratton to look into fixing his prototype while he went away to work independently on the problem from another angle.

Initially, all of them worked very well together. Bardeen would come up with new experiments and ways of testing, while Bratton would build whatever was required for the tests. Occasionally, Shockley and other staff would be called in for advice.

After some slight progress, in which Bardeen and Bratton achieved a barely amplified signal using a semiconductor, the pair became stuck and worked for almost a year with nothing more to show for it. Tensions began to build. Soon Bardeen and Shockley began butting heads, often.

In late 1947, there was a breakthrough in the research. On a whim, Bardeen decided to dip the whole apparatus in water and suddenly, the signal was drastically amplified. This was a shock to everyone. It was soon concluded that this phenomenon arose because the electrons in the semiconductor could now flow from its surface. After this, Bardeen and Bratton seemed to make constant progress, and it was only a matter of days before they eventually created a working transistor.

Excited, the pair called up Shockley to let him know that they had finally done it. Shockley was happy that the transistor had finally been created, but to the surprise of Bardeen and Bratton, he soon turned sour and angry at them. You see, Shockley was mad because the discovery of the transistor now could not be attributed to him.

Dismayed, Shockley turned recluse and worked frantically to try to do better. Looking at Bardeen and Bratton's designs, he realized that their invention had weaknesses: their device was fragile. How could they expect this transistor device to work in the real world if it broke all the time? Not to mention how hard it would be to mass-produce. Shockley set to work on building a more robust version of the transistor.

It would be less than a month before Shockley came up with a solution. All this time he told no one, especially not Bardeen and Bratton, who were still working on improving their original design.

Behind the backs of the other two, Shockley made contact with Bell Labs' lawyers to write a patent. He had one significant condition: only the name of Bill Shockley would be credited with the invention of the transistor. When Bardeen and Bratton heard about this, they felt betrayed. They had worked tirelessly for years to solve this problem, only to have it snatched away from them at the very end. In a twist, the lawyers decided to credit the patent to Bardeen and Bratton, as they determined Shockley wasn't involved in the core idea, or in the experiments with the transistor.

The Public Unveiling

On June 30, 1948, AT&T went public with the invention in New York. At the event, AT&T told the press that they were going to license the technology. In preparation for some publicity photos, Bardeen,

Bratton, and Shockley were asked to recreate the moment they discovered the working transistor for the camera.

What followed was a moment revealing of Shockley's true character: when asked to get in position for the photo, Shockley sat in Bratton's seat—taking center stage—while Bardeen and Bratton stood around him. Bratton would later write a letter to Shockley expressing his disappointment: "It would appear that the discovery of the transistor ruined the best research team I have had the privilege of working in." The resulting photo didn't matter much in the end, as the press didn't really see what the fuss was about. *Time* magazine put it in a modest section called "Science of the Week." At the time vacuum tubes were still (somehow!) seen as adequate by most: why replace them? Others saw the transistor as just a cute electrical toy.

This would change when two engineers from a bomb-damaged department store in Japan opened up a scientific publication and saw the transistor—a tiny device that could amplify signals. Akio Morita and Masaru Ibuka had just started a new company called Totsuko (later renamed Sony), and they realized that AT&T's invention could build the ultimate consumer product—a miniature, pocket-sized radio. While a sonic revolution was in the minds of the fledgling Sony company, other music-enabling technology was emerging.

CONSUMER MUSIC BECOMES MODERN

As the horror of war faded in the United States, and the sun was setting on the 1940s, the mood was once again a confident one. Consumer business was thriving once more, as factories turned back to producing civilian goods. This meant people had jobs, money to spend, and things to buy. Leisure items became a lucrative business. One such item was the long-playing record, or LP.

In 1948, the LP was introduced, opening the door to quality consumer music. It would be used as the standard format for vinyl albums, as LPs could hold many more songs than the traditional record. Interestingly, the word "album" originated from the need to store multiple sleeved discs in a book-like form, since the individual records couldn't hold enough audio. This changed with the LP, but the name "album" persisted and still does today.

The 1940s started with war and destruction, but left us with innovations such as early computers, the transistor, and the jet engine. These concepts were thought to be impossible by most people when the decade began, but great minds brought them into existence. The borders of the world had now been redrawn, but so had the possibilities for humanity. In the next chapter, we see these ideas form the bedrock for what would become the seeds of today. That bedrock was the 1950s.

PART 2

THE ACCELERATION

CHAPTER 8

The Seeds of Today
1950–1959

Strangely enough, the United States and the USSR had good relations directly after the war, but they didn't last. Winston Churchill made his famous "Iron Curtain speech" in 1946, and the mood of Americans began to shift toward fear of communism. Stalin's USSR was expanding across Eastern Europe, taking many countries in its wake. When China went red in 1949, a quarter of the world was now communist: it seemed it was only a matter of time before the ideology spread to the US.

By 1950, the horror of war was a fading memory. A deeper-seated, more subtle battle was taking place between two competing worldviews.

Despite this tense scene, a great shift was about to take place in most developed nations. For reasons touched on earlier, they began to prosper. Even countries that were heavily affected by the war saw a rebound in their economies. The boom was so massive that it would last all the way to the 1970s.

This newfound economic prosperity, coupled with technology developed from the desperation of war, set the stage for an absolute explosion in the rate of technological advance. With peacetime now in full swing, the 1950s would see everyday inventions such as roll-on deodorant, the credit card, Hula Hoops, power steering, automatic doors, the solar cell, lasers, and Barbie dolls. It also produced breakthrough moments in scientific and engineering brilliance that would change history forever; in essence, establishing the seeds of today.

Awe and Tragedy: The First Passenger Jet (1952)

It wasn't long after the war that the innovations of jet propulsion technology (from Frank Whittle, in the previous chapter) began to benefit society as a whole. In 1949, the world was shocked when the British unveiled the de Havilland Comet, the first passenger jet airliner and a strict secret until the day of its unveiling.

No other aircraft could match it. The Comet, dressed in silver metal, was beautiful to behold and could fly almost twice as high as the competition, arriving at its destination in half the time. Further, because it flew so high, there was less turbulence in its flight, meaning a smoother ride. In addition, instead of the noisy racket of a propeller engine, all that could be heard was an eerie ghostly whine from the four jet engines housed within the Comet's wings. It was almost the stuff of science fiction for civilians in 1949, and five years ahead of anything the Americans were dreaming of at the time. The de Havilland Comet would enter service in 1952, ushering in the age of the passenger jet plane.

As soon as the world saw how fast this jet could fly, interest, and eventually orders, began to mount. From France to India and Japan, even Britain's rivals, the United States, wanted a jet in their national airline fleets. With this event, the globe had suddenly shrunk. Things looked great for the de Havilland Comet, but suddenly strange accidents started to happen around the world. Upon takeoff, some Comets were crashing.

As it turned out, the design of the plane's wings would cause the aircraft to stall at low speed during takeoff. A stall happens when a plane no longer has enough lift to keep itself aloft, quickly losing altitude.

Jet aircraft were fussier when it came to climbing on takeoff, because there wasn't much air traveling over their wings until they reached a high speed. Propeller aircraft, on the other hand, had air from the propeller itself flowing over the wings to enable more control at lower speed. The problem was that early pilots were thinking of the Comet like the propeller-driven aircraft that had come before it and would fly it as such.

Things would only get worse, though. In 1954, two Comets seemingly just fell from the sky, only a few months apart. There were no survivors. All Comets were immediately grounded while the investigation was underway. At the same time, foreign orders for the aircraft ground to a halt. The reputation of the aircraft was getting hammered with each new development; what had once been the awe of the skies had now turned into a horror show.

In Britain, investigation teams worked to piece the planes back together in large hangars to establish what had caused the disasters. But how could they solve this mystery? There were no black boxes at the time to give data that could provide clues, or even a place to start.

What the investigators did next was a piece of engineering mastery. An intact Comet fuselage was placed inside a large sealed container, with the wings protruding. Water was poured into the container and then drained repeatedly. This was done to simulate the cycles of expansion and contraction of the fuselage that occur during multiple flights. What the experiments revealed was astonishing.

As the number of cycles increased to the equivalent of 5,000 three-hour flights, stress cracks began to appear in the corners of the windows. Over time, these cracks grew and eventually tore open. This would have been catastrophic during flight, as the internal air pressure of the fuselage was much greater than the outward pressure of the surrounding air. Basically, after a crack had formed, the plane would expand rapidly during flight (known as rapid decompression). To put it even more bluntly, the plane would explode.

Now, you may be wondering why I said the cracks formed from the corners of the windows. Aircraft windows are round, aren't they? Well, they weren't yet. You see, the Comet was fitted with square windows to give passengers a better view. This design choice concentrated a lot of the stress on the sharper corners of the windows, resulting in the formation of cracks. For this reason, all aircraft since then have been made with rounded windows, to stop this from ever happening again. The painful lessons learned from the Comet made future aviation safer for everyone.

Next Time There's a Crash, We'll At Least Know What Happened

The Comet crashes and ensuing investigations inspired an Australian researcher by the name of David Warren to invent a device, one which could record an aircraft's instruments and cockpit audio while in flight. He called it "the Flight Memory Unit." Today, it is known simply as the "black box" and has proven crucial to the successful investigations of countless air accidents, making air travel an ever-safer endeavor.

Surprisingly, the Comet would rise again to the top in the form of the Comet 4. It was a redesign with oval windows and a thicker metallic skin and was twice as large and twice as powerful as the original. The Comet would stay in operation up until 1997, when it was finally retired from service.

The Code of Life: DNA Structure (1953)

The 1950s would see the discovery of the DNA (deoxyribonucleic acid) structure. It's the chemical that encodes instructions for building and replicating almost all living things. It was formally discovered in 1953 by James Watson and Francis Crick (colleagues and close friends). This finding paved the way for genetic engineering, biotechnology, and molecular biology as a whole. That's what most people know about the discovery of the DNA molecule; as it turns out, though, there was another person who was involved in the world-changing discovery but was seemingly barred from all credit. Her name was Rosalind Franklin. To understand Franklin's contribution, we have to go back in time a little.

In 1946, Rosalind Franklin had been taught how to use X-ray diffraction by Jacques Mering, a French crystallographer. She would go on to pioneer the use of X-rays to yield images of complex forms of matter. Franklin went on to research at King's College in London, taking her X-ray expertise with her.

In 1951, at King's College, Rosalind Franklin and Maurice Wilkins (who became Franklin's senior partner) were studying DNA. Wilkins and Franklin used X-ray diffraction as their main tool. Beaming X-rays through the molecule yielded a shadow picture of the molecule's structure, as a result of how the X-rays bounced off the molecule's component parts. Franklin's knowledge was used to create a method of "seeing" the DNA molecules.

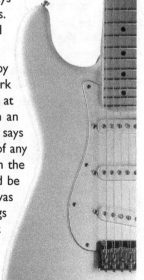

Rosalind was shy and often patronized by colleagues; she had to carry out much of her work alone because of this. She was cautious and skilled at her craft, refining X-ray photo machines to such an extent that renowned UK scientist John Desmond says she produced "the most beautiful X-ray photos of any substance taken." One day, in January of 1953, in the lab, a clear X-ray image of the double helix could be seen. Maurice Wilkins, seeing Franklin's work, was astonished, and showed some of Franklin's findings to molecular biologist James Watson without her knowledge. Watson and Wilkins had met a few years earlier in Italy and had previously discussed the use of X-ray diffraction to deduce the structure of the DNA molecule.

Franklin's X-ray image (known as "Exposure 51") caused Watson to have a visceral reaction. Watson is reported to have said: "The instant I saw the picture, my mouth fell open and my pulse began to race." Shortly after, Watson teamed up with Crick and made a crucial advance.

They proposed that the DNA molecule was made up of two chains of molecules called nucleotides, arranged parallel to each other to form a "double helix," like a spiral staircase. This structure, published in the April 1953 issue of *Nature*, explained how the DNA molecule could replicate itself during cell division, enabling organisms to reproduce with amazing accuracy, except for occasional mutations.

For their work, Watson, Crick, and Wilkins each received the Nobel Prize in 1962. Despite her contribution to the discovery of DNA's helical structure, Rosalind Franklin was not named a prize winner: she had died of cancer four years earlier, at the age of thirty-seven. Forgotten by history and dying of cancer at a young age, she is strangely and sadly reminiscent of Ada Lovelace in the chapter before last.

Polio is Defeated: The Polio Vaccine (1953)

Poliomyelitis (otherwise known as polio) is a viral infection that alters the body's host cells in order to replicate itself. The virus has effects ranging from a sore throat to fever; however, when it enters the spinal cord and nervous system, muscle wasting and paralysis can occur. Although the latter is rare, the scary thing is that loss of the ability to move can happen in as little time as a few days or hours. On top of that, polio was transmitted extremely easily from person to person.

In the early twentieth century, as highly populated cities became more common, polio epidemics began to occur for the first time in history. In 1916, over 2,000 people died from the infection in New York City alone, and many more were left paralyzed. It was a time of great uncertainty. With each wave of polio outbreaks, people would flee the cities and public gatherings would stop. In 1952, the worst outbreak in the US occurred: almost 60,000 cases were reported, with 3,000 dead and 21,000 disabled. To this day, polio has no cure; but, on the very bright side, there is a vaccine.

Dr. Jonas Salk is the one we have to thank for that. During World War II, Salk aided in the development of flu vaccines and became the lead researcher at the University of Pittsburgh. In 1948, Salk would be awarded a large research grant to develop a polio vaccine, since it was such a pressing health issue of the day. Within just two years, Salk had an early version of the vaccine.

Here's how it worked: Dr. Salk devised a way to kill multiple strains of the polio virus and injected the inactivated viruses into the bloodstream of a healthy person. As a defense, the patient's immune system automatically created antibodies that would fight off infection, without the patient ever falling ill to the disease. This technique was tried in the 1930s by Maurice Brodie but proved unsuccessful. Interestingly, many scientists of the day thought that Salk's method wouldn't work. Their incorrect view was that a dead virus wouldn't encourage a strong enough immune response from the body. They thought that a weak, live virus was needed do the trick. Salk, on the other hand, didn't want to take that risk.

In a demonstration of complete faith in his work, the doctor conducted the very first human trials on himself, his family, and later, former polio patients. By 1953, it was clear that his method was working. On March 23, 1953, Salk published his findings in the *Journal of the American Medical Association*, and two days later appeared on CBS radio to talk about what he had done. America and the world collectively breathed a sigh of relief and hope.

In 1954, clinical trials began on two million American schoolchildren. In April 1955, it was announced that the vaccine was effective and safe, and a nationwide effort began. Today, polio has virtually been eradicated worldwide, with only 113 cases out of seven billion people being reported in 2017.

Just imagine if polio continued to ravage humanity with the densely populated cities of today: it would result in pandemonium. Dr. Salk was truly the MVP of medicine in the 1950s.

The Most Iconic Instrument Ever Made: The Fender Stratocaster (1954)

Like most guitarists, I can always tell the sound of a Fender, especially a Stratocaster. It has a unique character. The only way I can describe the sound is as "expressive": a sound allowing artists to communicate all kinds of subtle emotions. Because of the positioning of the Stratocaster's pickups (the magnetic devices that convert string vibrations into audio signals), the guitar has a versatile tone that can be used in most genres of music: from country, to rock, to soul, jazz, and punk. The Fender Stratocaster has become the most legendary instrument of all time. From Jimi Hendrix to John Mayer, to Tom DeLonge of Blink-182—they're seen everywhere. But where did the Stratocaster come from?

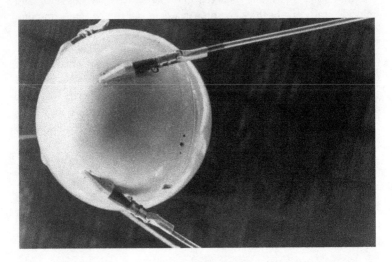

During World War II, Leo Fender was running a radio repair shop in Fullerton, California. After some time making amplifiers and experimental electric guitars, Fender created his first electric guitar— the Esquire. In 1951, this was followed by the Telecaster, with its unique snappy and "quacky" sound. Building on the design of the Telecaster, in 1954, Freddie Tavares, who worked with Fender and had an eye for radial curves, gave the Stratocaster its radical, timeless, flowing shape.

Surprisingly, Leo Fender wasn't a guitar player; he was an engineer. But under his management, the company managed to produce a range of guitars and bass models so popular, their sound permanently influenced music as a whole.

The Video Cassette Recorder (1956)

The tape recording of television signals became feasible after World War II, but the equipment of the time was pushed to the limit in doing so [tapes had to run at 240 inches (610 cm) per second to obtain the high-frequency response needed for television signals]. Charles Ginsburg thought he could do better. His new machine could do the same task, without the machine running so fast that it emitted proverbial smoke.

This video tape recorder (called the Ampex VRX-1000) was released in 1956 and changed television forever. Instead of live broadcasts, programs could now be shot and more easily edited after the fact.

Looking Inside: The Ultrasound (1956)

Ian Donald was a professor of midwifery at Glasgow University. While at a shipyard in Glasgow, he noticed that the workers used sonography to look for flaws within the thick surfaces and structures of ships. Sonography works by emitting sound waves above the range of human hearing against a surface and measuring the waves that are reflected back. These reflected echo-waves are converted into an electrical signal, which is interpreted as an image. In 1956, Donald and an engineer, Tom Brown, decided to see if this could be used in the medical field. Early tests proved the technology to be feasible, and just two years later, they would publish their findings. The medical tool of the ultrasound was officially...born (sorry, that one just lined right up).

First Programmable Robots (1957)

The story of the first robot starts in an unexpected place. Not a shed, or workshop...but a cocktail party in 1956. Two of the attendees were George Devol, an inventor, and Joseph Engelberger, a businessman—both fans of science fiction. Obsessed with the idea of a true robot, Engelberger was delighted when Devol told him about his recent creation, the "Programmed Article Transfer." In Devol's patent, it was described as "an invention [that] makes available for the first time a more or less general-purpose machine, [which] has universal application to a vast diversity of applications where cyclic digital control is desired." It was basically a controllable robotic arm that was programmable.

Devol had been part of many novel inventions, including one of the first microwave oven products, the "Speedy Weeny" (which automatically cooked and dispensed hotdogs—an essential product for anyone).

Devol and Engelberger together produced the Unimate (short for Universal Automation). It was the first truly modern, digitally-operated, programmable, teachable robot.

A commercial version of the robot was released in 1959 and quickly found its way into the automotive manufacturing plant of General Motors. Soon, auto manufacturers in the United States and Japan were reaping the rewards of automation, thanks to the Unimate. In 1966, Engelberger appeared on the *Tonight Show* with Johnny Carson to demonstrate the robot, which was capable of performing a wide range of tasks such as pouring drinks and handling various objects. The public's imagination went wild.

Today, robotic arms are an integral part of many industries worldwide.

The Clueless Baby—First Digital Picture (1957)

In 1957, Russell Kirsch was a scientist at the National Bureau of Standards and had access to the Standards Electronic Automatic Computer (SEAC), one of the only programmable computers in the United States. It took up a large portion of a room and had a clock speed of one megahertz (1 million switches per second), about 2,800 times slower than a smartphone. Kirsch, who led the imaging research team for the computer, decided to see if computers had the ability to reproduce pictures. Kirsch scanned in a five-centimeter-square photo of his infant son, which was turned into binary code (1s and 0s). The result was a 176-pixel, black-and-white image of a baby, clueless that it had just made history. The digital photo and digital camera itself would come two decades later. It goes without saying that digital photography is the form that essentially all photos exist in today.

A Beach Ball in Space—First Man-Made Object in Orbit (1957)

On October 4, 1957, mankind would send objects into space for the first time in history. The Soviet satellite, Sputnik 1, became the first artificial object put into orbit. It traveled around the earth once every 96 minutes. Although Sputnik was only the size of a beach ball, and weighed just 83 kg, the world was stunned at the event. Interestingly, the Soviets had actually planned to launch a much bigger, 1,400-kilogram spacecraft with many instruments on board. Unfortunately for the Soviets, the development of the larger craft was so slow that those managing the project grew anxious. They feared that, if they didn't hurry up, the Americans would beat them to the punch. In the end, a metallic silver beach ball had to do for mankind's first venture into space. Sputnik was a brief mission, sending out a radio signal for only twenty-two days before burning up in the atmosphere in early 1958.

The launch of Sputnik was a significant event. It prompted the foundation of NASA and DARPA by the Americans, marking the beginning of the space race. As we'll see in the next chapter, Sputnik would result in ARPANET, the start of our current internet. Sputnik also indirectly accelerated the invention of something that would become integral to the internet, and something we all can't live without today: the digital modem.

You'll Appreciate Your Internet After This: The Digital Modem (1958)

The origins of the modem came from the need to connect typing machines via phone lines. In 1943, IBM would use this technology to transmit punch-card data at 25 bits per second (or about three letters per second)—slow data transfer speeds to say the least.

A year after the launch of Sputnik (pictured above), the United States air defense systems were due for an upgrade. The systems needed a way for computer terminals in various locations to communicate information. After looking at their

Tiny Transistor VS
Bulky Vacuum Tube

options, they contracted AT&T for the job. In 1958, the company came up with a digital version of the modem to connect terminals across the United States and Canada.

These modems operated at 110 bits per second.

For some perspective:

- An average webpage (500 KB) would have taken fifteen hours to load.

- A high-quality mp3 download would be all yours in just under two weeks.

- And a five-minute, 1080p YouTube video download (at 12 MB per minute) would have taken around two and a half months!

Imagine clicking on a video, then coming back to watch it two and a half months later. Knowing this, we have no reason to complain about our internet speeds.

The first commercial modem was released in 1962 by AT&T, at three times the speed of the 1958 version. At least it was an improvement.

Making Things Interesting—The First Graphical Video Game (1958)

For many of us, video games have always been around: from arcade games like *Pong* and *Pac-Man*, to more recent forms via consoles and virtual reality. What many don't realize is that the video game's history stretches back to the late 1950s, as the brainchild of physicist William Higinbotham.

During the war, Higinbotham worked on cathode ray tube displays used for radar systems, as well as for electronic timing systems in the atomic bomb. After the war, he got a job at Brookhaven National Lab's instrumentation group.

The Brookhaven firm had an annual open house where thousands of members of the public would come and tour the lab. Higinbotham was put in charge of creating an exhibit that would showcase some of the instrumentation the company had to offer. As he was thinking of what to do, he realized that a lot of the items on display were very... well, boring. It would be like people coming all the way to watch paint dry, but less fun. He had an idea on how to liven up the place: what about using a game that people could interact with and play? It would give a strong message that the lab's scientific endeavors had relevance for society.

With this idea in mind, Higinbotham rummaged around for parts he could use to build an interactive game. He came across an analog computer that could display the curved trajectories of objects, such as a bouncing ball on an oscilloscope. Because of his experience with radar displays during the war, it only took the physicist a few days to put a game together. The result was a crude but playable side-on view of a tennis court. It consisted of two lines: one for the ground and one for the net. The game's name? *Tennis for Two*.

The controls were simple—a twistable knob to adjust the angle at which the player was about to hit the ball, and a push button to fire the shot.

In October of 1958, the first group of visitors came in to see what the Brookhaven lab company had on display. People were captivated by *Tennis for Two*, and small crowds began to form around the game. Although popular, Higinbotham didn't think much of it, having no idea how popular the concept of a video game would later become. He stated:

"It never occurred to me that I was doing anything very exciting. The long line of people, I thought, was not because this was so great, but because all the rest of the things were so dull."

Today, the video game industry makes around $110 billion in revenue annually. You're a modest guy, Higinbotham, very modest.

The Society Changers

As you can see, there was a flood of groundbreaking inventions that came from the 1950s, happening almost on a yearly basis. To finish off this chapter, we'll take a look at three society-changing innovations: portable music, the integrated circuit, and television.

A LEGEND IS BORN: SONY

In the last chapter, the transistor had been created by the AT&T/Bell Labs trio but was largely ignored until Masaru Ibuka and Akio Morita of the newly-founded Sony stumbled upon it. They recognized the technology for its ability to amplify sound. Fascinated, the duo decided to improve the device and make it more affordable for mass production. Sony had been trying to sell tape recorders locally in Japan, but were struggling, as post-war Japan was still largely in ruins. An obvious first application for the transistor was enabling a much smaller radio, as mentioned earlier.

Vacuum tube (pre-transistor) radios of the time were usually wooden cabinet sets that took up a lot of space in a living room. A transistorized version could use much less power and be a fraction of the size. However, Ibuka and Morita weren't the only ones with this idea. Recognition of the technology's abilities had begun to increase, and American electronics manufacturer Regency beat Sony to the punch. They released the TR-01 portable transistor radio in 1954. It cost three times as much as a vacuum tube radio at $50 ($460 today). It was a sonic revolution. Consumers no longer had to wait for their radio to warm up, and it was so light and small they could carry their music around with them anywhere.

Even though these radios were extremely popular, due to their high production cost, they still proved unprofitable for the companies that made them. In addition, the US military would pay $100 ($920 today) for just one high-quality transistor component. It was a no-brainer. It made much more sense for the American companies to leave the consumer market and go into the much more lucrative military industry. As a side effect, the transistor radio market was left wide open for someone who could figure out how to make transistor radios cheaply enough. Sony was up to the job. Morita and Ibuka worked hard on improving the economic feasibility of transistor technology. Just one year after the Regency TR-01, Sony had an affordable version of a transistor radio.

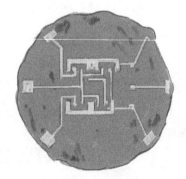

In April of 1955, Morita traveled to the United States to sell his radio. Surprisingly, it met with a "meh" response by most prospective business partners. This was so until Sony approached a company called Bulova, which was excited about the idea. Bulova was a watch manufacturer, and it understood what Sony was trying to do. The

Early integrated circuit (connected by chemical etching instead of individual wires)

company put down an order for 100,000 units, which translated into more money than Sony was worth at the time. The only catch? Bulova wanted its name on the product. Morita said no and walked away from the deal. Morita would later say this was the most important business decision he'd ever made.

Morita went back to Japan and back to the drawing board. Sony knew it was close to a commercial hit. It just needed something enticing to push people into making a purchase. The answer to this issue turned out to be a brilliant piece of marketing: How about calling it a radio that "fits in your pocket"? People would really understand how revolutionary that is! The problem was that the radio most certainly didn't fit in the average-sized shirt pocket.

The solution? Alter the shirt pockets of the sales team so they were big enough to fit the radio. It worked. Soon the United States store owners were lining up to buy the Sony radios (now named the TR-55s). Within five short years, the Sony company grew from seven employees to over 500. And with that the legend of Sony was born, along with a new era for portable music long before the Walkman, iPod, or mobile phone.

For the first time, kids could listen to music privately, out of earshot of their parents. This newfound freedom sparked a change in the type of music that was broadcast on certain stations. It became more edgy, daring, and risqué. This new sound would be the beginnings of rock and roll. This was the first direct impact the transistor had on society, but there would be many more to come.

The Gateway to Modern Computing: The Integrated Circuit

Transistors were proving useful in radios, but their use in functioning computers was limited. If enough transistors could be packed into a sufficiently small space, there could be serious computing power. The problem was the limit to how small you could make a transistor of that era. They had to be connected by wires. As you can imagine, these connections could only be so tiny before becoming unreliable. Scientists wanted to condense a whole circuit—the transistors, the wires, everything else they needed—into a single step. If they could create a miniature circuit, all the parts could be made much smaller, and also mass-produced.

The answer was the integrated circuit. The idea was to use a chemical etching process, instead of wires, to create the transistor connections.

It was the next major leap in the path toward our current world. Integrated circuits are found in virtually all electronics today.

The story of the integrated circuit starts once again with the AT&T/ Bell Labs dream team. By 1950, the team was starting to crumble, as John Bardeen and Walter Bratton became increasingly frustrated with William Shockley. In 1951, Bardeen decided to call it quits. He couldn't work under Shockley anymore, and with that, the dream team was over.

By 1955, Shockley's reputation at Bell Labs had gone from bad to worse. It was clear he wasn't going any further in the company. So, Shockley aimed to start his own company, with the goal of creating a new kind of improved transistor.

Unsurprisingly, nobody wanted to work with him. So, with his head down and his proverbial tail between his legs, Shockley closed up shop and moved from the East Coast out west to Palo Alto, California. Here he hoped to recruit some new team members who knew nothing of his controversial history.

At the time, California was a strange place to set up a technology company, as most firms such as IBM (we'll get to them later), Raytheon, and General Electric were based on the East Coast. Shockley's motivation to move was partly due to family, but more truly it was due to the fact that Stanford University was offering cheap land to entice technology companies to go west.

Eventually, in 1957, William Shockley would set up the humbly named "Shockley Semiconductor" Lab. As rotten as Shockley's personality was, he did have a great ability to source talent. He managed to recruit a number of young men who were unaware of his turbulent past. The new team that Shockley cobbled together happened to include eight of the greatest engineers and scientists the United States had to offer. They would be known as the Traitorous Eight.

THE TRAITOROUS EIGHT

Shockley Semiconductor Lab brought together a pool of talent that would become the foundation of what we now know as Silicon Valley. Those men were: Sheldon Roberts, Eugene Kleiner, Victor Grinich, Jay Last, Julius Blank, Robert Noyce, Gordon Moore, and Jean Hoerni.

In June of 1957, these eight men held a secret meeting in a hotel in San Francisco. During their time working for Shockley, they had all noticed how his personality had become increasingly paranoid. Much like at AT&T, Shockley was afraid that he wouldn't receive credit for the team's inventions. Sitting in the hotel room, the men passed around a dollar bill, each taking their turn to sign it. That bill represented a pact to leave Shockley Semiconductor and start their own company. This plan earned them the name of "The Traitorous Eight."

This new company was to be under the reluctant leadership of Robert Noyce, a brilliant MIT researcher. Noyce was hesitant to take on such an ambitious project, as he had a young family, and start-ups were a huge risk in the 1950s. But Noyce bit the bullet and signed the dollar bill. With that, the traitorous eight had a new spinoff company: Fairchild Semiconductor.

Fairchild Semiconductor

That day in the San Francisco hotel would later be called one of the top ten days that changed history by the *New York Times*. Fairchild was the essence of what Silicon Valley grew to stand for: risk over stability, innovation over tradition, rapid experimentation over slow growth. After securing financial backing, the group at Fairchild came together to produce the first practical integrated circuit, using the silicon semiconductor. The invention made it possible to tightly pack transistors but allowed them to function independently despite being made from a single piece. The integrated circuit (which is itself made from transistor components) is the building block of all chips. It caused a revolution in electronics, opening the door for today's technology.

Steve Jobs in 2005 would say that working in Silicon Valley was a bit like running a relay race, and Fairchild Semiconductors were the ones who passed the baton to him to bring in the era of personal computing.

The treacherous eight at Fairchild made it possible for a modern smartphone to be more than 120 million times more powerful than the computers that sent man to the moon.

Moore's Law

One member of the Fairchild team, Gordon Moore, made a prediction in 1965: "The 'complexity number' of transistors (processing power), on a chip of a computer, will double once every eighteen months." This observation would be known as Moore's Law. This law wasn't based on any calculation, but more on what was expected of the industry. Surprisingly, since its inception, Moore's law has held true.

Gordon Moore would also become famous for something else. In 1968, he and Robert Noyce became the co-founders of a little company whose name you may have heard of: Intel.

In the end, there was a silver lining for the Bell Labs dream team. In 1956, William Shockley, John Bardeen, and Walter Bratton would each receive the Nobel Prize for physics.

Television Enters the Picture

During the 1940s, television was struggling to make its mark. In Britain, it was seen as a novelty for the rich, with a set costing roughly seven times the weekly wage. Back in the United States, all programming was filmed live, with all three networks, NBC, CBS, and ABC, filming out of New York. Production was simplistic by today's standards—point, shoot, and ship the 16-mm film around the country to be broadcast later.

By 1950, things had begun to change. An estimated 4.4 million American families owned a TV. AT&T had created a web of coaxial cables spanning the country. TV signals could now reach the most rural areas of America and soon, for the first time in history, the country would not just hear but also see the same things at the same time. News, popular TV shows, and sports became a shared national experience. In Britain, Europe, and Japan, a similar revolution was happening.

Color TV would follow in 1953. Coincidentally, that was the same year TV dinners were invented. It looked like the medium was here to stay.

A typical TV looked an awful lot like a box. It was usually housed in a large wooden cabinet and had a rounded screen, which was tiny in comparison to the fifty-inch flat screens of today.

Movie attendance dropped, as did radio listenership. And while the '50s had its share of classic shows, it wasn't until the '60s that the shared experience of television would really begin to shape culture, with world-shaking events such as the Vietnam War and the moon landing being broadcast globally. TV changed the way we saw the world, as well as the speed at which we saw it. The assassination of John F. Kennedy, for example, spread to 90 percent of America in less than an hour, largely through live broadcasts. For better or worse, the world was now connected.

CHAPTER 9

Peace, Love, and...Computers?
1960–1969

The 1960s brought on new ways of thinking, innovative ideas, radical ways of dressing, and new forms of music—and all of it started with the baby boomers.

The baby boomers of the post-war era were coming of age in the '60s. The strict rules of the '50s gave way to a newfound sense of rebellion among the young. So dramatic was the shift that a new word was popularized for the group: "teenagers." And the economy was ready to support them. The youth of the '60s found themselves in a surprisingly comfortable economic situation. In fact, people under twenty-one were paid ten times what the same demographic were paid before the war. The spending habits of America's teens began to drive the economy. Many companies founded during this decade reflected the affluence of the time. Domino's Pizza, Weight Watchers, and Subway all came into existence in the '60s. Fashion began to reflect personality, as starched collars and letterman jackets gave way to wild new styles. Artists such as Bob Dylan, the Rolling Stones and the Beatles became the sounds of a generation.

Between 1954 and 1965, the number of young people attending secondary school in the US doubled. The same was happening in Japan and the UK. For many, it was the first time someone in their family went to university.

In the world of technology, meanwhile, innovation with transistors and integrated circuits caused computers to shrink to a usable size. Silicon Valley came of age and venture capitalism was born. The Concorde became the world's first supersonic airliner, the cassette tape was born, and the seeds of the compact disc were planted. The decade was full of revolution, and revolutionary ideas.

THE FIRST MAN IN SPACE (1961)

The world was shocked when the Russians sent Sputnik into space, but the boys from the USSR weren't finished. On April 12, 1961, aboard

the spacecraft *Vostok 1*, twenty-seven-year-old Soviet cosmonaut Yuri
Gagarin became the first human to travel into space.

Vostok 1 orbited the planet for just eighty-nine minutes at a maximum altitude of 203 miles and was guided entirely by an automatic control system. It was basically a cannonball large enough for one person. Gagarin wasn't very talkative during the ordeal; his only words were: "Flight is proceeding normally; I am well." This may have been in part due to the fact that Gagarin was traveling at around 29,000 km an hour (17,500 miles per hour). That's 8 km (or 5 miles) per second!

Like Charles Lindbergh with his first solo transatlantic flight a few decades before, Gagarin became an instant worldwide celebrity. In Russia, he ascended to legendary status. Monuments were raised, and street names were changed throughout the Soviet Union in his honor.

The United States weren't just being beaten to the punch this time, they were getting punched square in the face. To add insult to injury, the first US space flight was scheduled for May of 1961, a mere one month after Gagarin. Then, in August of 1961, the Americans were beaten again with the flight of cosmonaut Gherman Titov in *Vostok 2*. Titov made seventeen orbits of the earth and spent more than twenty-five hours in space.

America's first manned mission was delayed until February 1962, when astronaut John Glenn made three orbits in *Friendship 7*. Understanding that he needed to boost his country's confidence, President Kennedy declared that the US would reach the moon by the end of the decade.

Meanwhile, in the USSR, the Soviet conquest of space was viewed as evidence of the supremacy of communism over capitalism. However, only insiders who worked on the Sputnik and the Vostok programs knew that the successes were actually due to the brilliance of one man, a new thinker in every sense of the word: Sergei Pavlovich Korolev. Chief Designer Korolev was unknown in the West, and even to the Russian public, until his death in 1966.

Korolev helped launch the first Soviet liquid-fueled rocket in 1933. Five years later, during one of Joseph Stalin's paranoid purges, Korolev and his colleagues were put on trial. Convicted of treason and sabotage, Korolev was sentenced to ten years in a labor camp. As World War II intensified, the Soviet leaders started to realize that German rocket technology was a real threat. Knowing of Korolev's talent, the government put him to work from prison. I'd imagine he would have been less than thrilled.

In 1945, when the war was over, Korolev was sent to Germany to learn about Nazi rocketry. His mission was to study the V-2 rocket, which had heavily damaged Britain during the war. The Soviets weren't the only ones stealing German notes for the rocket exam—the Americans were at it too. The United States captured the V-2 rocket's designer and eventually made him head of the US space program.

The Soviets went further and managed to get their hands on rockets, blueprints, and a few German V-2 technicians. By 1954, Korolev had built a rocket that could carry a five-ton nuclear warhead. In 1957, the Russians launched the first intercontinental ballistic missile.

But Korolev had his eyes set on bigger things, which brings us back to the launch of the metallic beach ball known as *Sputnik 1*. It was the first Soviet victory of the space race, but amazingly Korolev was still in prison! He would finally be released a short time later, and with Korolev at the helm, the Soviet space program was unmatched in the early 1960s. The USSR's list of firsts into space include: first animal in orbit, first large scientific satellite, first man, first woman, first three men, first spacewalk, first spacecraft to impact the moon, first to orbit the moon, first to impact Venus, and first craft to soft-land on the moon.

Despite all his success, Korolev was known only by the mysterious title of "Chief Designer." Korolev died in 1966. Upon his death, his identity was finally revealed to the world, and he was awarded a burial in the Kremlin wall as a hero of the Soviet Union. I'm sure he would have appreciated this kind of honor a little more when he was still alive.

For most countries, having a man lead the space program from prison would be as strange as it gets; for Russia, it was just the tip of the iceberg.

The Curious Case of the First Spaceman

On March 27, 1968, seven years after his trip to space, Yuri Gagarin got up early in the morning and headed to Chkalovsky airport. His purpose was to re-train as a fighter pilot after his time being a cosmonaut. The plan was to have Gagarin run a few test missions in his Russian MiG-15, but things didn't go according to plan.

The weather was poor at the airport, with the wind and rain making their presence known. When Gagarin found he'd forgotten his ID that morning, he told those around him that this was a bad omen. Despite the weather and the bad juju, Gagarin took off without incident. After he'd completed his exercises, he radioed back to base that he would be returning shortly.

That was the last time anyone heard from him.

A couple of hours later, a flight rescue team took off on a search mission. As they were searching, they could see the plane's burning wreckage from the air. There was hope that the former cosmonaut might have ejected before impact. These hopes were dashed when Gagarin's body was found the next day. An extensive investigation by the USSR proved inconclusive. One prevalent theory was that Gagarin swerved to avoid an object (a bird or weather balloon) and lost control of the aircraft. Naturally, this being the Soviet Union, very little information was released to the public and the conspiracy theories started to grow.

Some say Gagarin was drunk and decided to take a joy flight. It's not unimaginable, as Gagarin had a history with the bottle. The young cosmonaut had taken his meteoric rise to international fame fairly hard. As a former village boy, he wasn't prepared to be an icon of inspiration. He would receive thousands of letters from citizens across the USSR telling their disturbing stories and asking for help. It was then that he began seeing the dark underbelly of the Soviet Union. As time went on, Gagarin began to slip into alcoholism and other risky behaviors.

Other theories suggest that he was a secret agent from the CIA and was poisoned. There was also a rumor that he had actually survived the plane crash and had gone into hiding. Still others said that he was trying to avoid a UFO when he crashed. So, what really happened? Will we ever know?

Well, you're in luck, because in 2013, the cause of the crash was finally determined. As it turned out, another much larger plane (a Sukhoi Su-15) came into the airspace of Gagarin's MiG-15, causing him to take evasive action and crash. In the end, it was a run-of-the-mill accident, ending in tragedy. The Soviet authorities and air traffic control were embarrassed about their incompetence, so the true cause wasn't released.

The First Digital Video Game (1961)

You have heard how William Higinbotham made his exhibit more exciting by creating *Tennis for Two*. Although this marked the beginning of the video game, the media form was yet to have its breakthrough.

As we noted, the decreasing size of transistors led to a reduction in the size of computers over the years. Because these newer computers no longer filled an entire room, they were called "minicomputers." Minicomputers weren't just small, they also had broader uses than their predecessors. No longer were they just tools for calculation and record-keeping.

Despite all this progress in the computing space, there was still no such thing as a video game industry. Almost all games had been developed on a single machine for a specific purpose. A software game that could be run on multiple machines was not yet realized.

This all changed in 1961 with MIT's acquisition of the Digital Equipment Corporation's Programmed Data Processor-1 (PDP-1) minicomputer. The PDP-1 used a vector display system. Even though it had no CPU, the transistors still managed a 5-megahertz (five million switches per second) processing speed, which is about 500 times slower in clock speed than a modern smartphone, but still powerful for the day. The

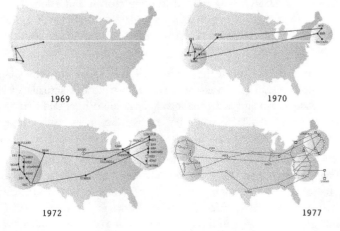

1969   1970

1972   1977

Growth of ARPANET

PDP-1 was made from assembled boards of transistors called "System Building Blocks."

Because of the PDP-1's small size and high speed, MIT students and employees enjoyed writing non-academic programs on it whenever it wasn't in use. In 1961–1962, Harvard and MIT employees Martin Graetz, Steve Russell, and Wayne Wiitanen created the game *Spacewar!* on the PDP-1.

The two-player game had players engaged in a dogfight in space. One of the aims of the game design was to utilize the PDP-1 to its fullest potential. The result was a surprisingly smooth game, given the limitations of the hardware.

Spacewar! was copied to minicomputers in several other American universities, making it the first video game to be available at more than one location. However, the game wasn't exactly widespread. The PDP-1 cost $120,000, which is almost $1 million in today's currency, so only fifty-five were ever sold.

In the years following the spread of *Spacewar!*, programmers at universities began developing and distributing their own games. We'll pick this story up in the next chapter, where we'll see how video games burst into the mainstream.

The Seeds of the Internet and the World Wide Web (1962)

The seeds of the internet and web were actually planted before 1962. You may not have heard his name, but Vannevar Bush—World War II military science leader, engineer, and thinker—is the man responsible for the contemporary information age. Bush's recognition has only been more widely recognized in the 2010s, as the internet became more prolific. Among many other things, Vannevar Bush founded the Raytheon company, and led 30,000 scientists, engineers, and high-ranking military men in research during World War II. Bush had immense power and answered

Sketchpad was the first graphical interactive computer program

only to the President himself. At one point, around two-thirds of all American physicists were working for him.

Vannevar Bush: The Father of the Information Age

Vannevar Bush, in my opinion, is one of the most influential figures in the entire book. His ideas inspired many technologies central to our modern life, as those that came after him built upon them. During his efforts in coordinating those 30,000 academics, he quickly realized that there was no way to catalogue all their research information. It would be a tragedy if great post-war scientific discoveries were lost in a sea of meaningless publications. In a 1945 article titled *As We May Think*, Bush envisioned a machine that would "link" information the way our brains do: by association.

This machine was called a Memex (short for MEMory EXtension). The Memex was to be a universal library—something bearing a striking resemblance a desktop PC. The research information (text, images) was to be printed on microfilm and magnified onto a display. If a researcher wanted to find out more about a particular topic, he could request to have linked information (on microfilm) sent to him.

For example, a science researcher could use a Memex to read a study and request all the associated (linked) images and studies for viewing on his Memex.

The information wouldn't be transferred between Memexes digitally, but by mail (this was 1945, after all). This is kind of like clicking on a Wikipedia article and waiting for the information on the linked webpage to be mailed to your door.

Let's break down the significance of this idea in today's terms:

- The mail service would be the **internet**.
- The ability to request information on a Memex is the **web**.
- The concept of linking information is what we now call **hyperlinks** (the way in which we click from one place to another on the web today).

Bush believed that such a device would cause an information explosion and subsequently a knowledge explosion. I can't say he was wrong.

In essence, Bush had envisioned the World Wide Web, forty-five years before it was developed in 1990. He did not live to witness

NEW THINKING

Memex Machine

it, but every one of his visions came true, to a greater degree than even he could have expected. Bush was indeed one of the greatest visionaries of the last century.

Drawing Up a Blueprint

Although Bush had envisioned the idea of the internet, it was only in the 1960s that technology advanced far enough to start building it.

In August of 1962, using Bush's work as a basis, psychologist and behavioral scientist J. C. R. Licklider (known as "Lick") would lay out the blueprints of the internet. Lick worked for the US Department of Defense as Director of Behavioral Sciences, Command & Control Research at the Advanced Research Projects Agency (or ARPA), which had been created one year after the launch of Sputnik.

In the early 1960s, the ARPA still dealt with all research by hand, which was not ideal. For example, for MIT and Dartmouth to transfer data, someone had to hop on a train, physically transport punch cards, and wait two to three hours for a printout before returning. Talk about slow download speeds.

Faster knowledge transfer was paramount, especially as the Russians had just been to space...twice. Lick saw a solution: computers needed to talk to each other directly. In 1962, he wrote "A Man and Computer Symbiosis." The book expressed the idea that machine cognition would become independent of human knowledge. This involved the concept of a "Thinking Centre"—a precondition for the development of networks.

In 1964, he set to work on building his connected vision and, in 1967, he described a "library of the future" that could have all of the world's books online, so anyone could access them.

He also envisioned online collaborations. People could communicate with each other as well as access shared information from anywhere in the world. Lick explained:

"The unique thing about computer communication networks is that not only can co-workers (who are geographically distributed) stay in touch with each other, but also with the information base in which they work all the time. [This means] that finished blueprints don't have to be printed out and shared all around the country but appear on everybody's scopes (screen). This is obviously going to make a huge difference in how we plan, organize, and exercise almost anything."

Lick would also imagine the banking system:

"If we get into a mode in which everything is handled electronically, and your only identification is on a little plastic thing that you stick into machinery, then I can imagine that [the supplier would] want to get that settled up with your bank account…and that would require a network."

As you can see, Lick didn't just have a rough idea of how the future was going to work: he completely understood what the end product should look like.

Even though Lick left ARPA in 1964, he passed his vision on to those next in line to take the helm of ARPA. The early internet became a reality in 1969 with introduction of ARPANET. ARPANET allowed communication between UCLA and Stanford computer terminals and was the direct precursor to the modern internet.

The first ever message sent over the network was "Lo-." Now you might be thinking: that's a funny word, why choose that? Well, the first message was supposed to be "Login," but the system crashed before reaching the "g." Nevertheless, this moment marked the beginning of a new age. ARPANET grew over the next decade, connecting computers around America.

It's incredible to see how quickly technology had begun to move. In the late '50s, computers were still cold, calculating number machines. But by 1963, twenty-five-year-old MIT graduate Ivan Sutherland had read Vannever Bush's 1945 essay "As We May Think." In the essay, Bush describes the "Memex" personal knowledge machine. Inspired by this idea Sutherland asked: what if you could "talk" to a computer? Not with text or speech, but with drawing. The result was

LSD

Serotonin

Sketchpad, an interactive drawing program in which the user drew with a light-pen while pressing switches alongside the screen. Sketchpad was the first use of computer graphics and brought about a new category of software: Computer Aided Drafting (CAD). Amazingly, it all ran on a 1958 computer with roughly 272 KB of memory and a 9-inch oscilloscope screen (a screen that shows signals, e.g. a heart monitor). It was a revolutionary idea, paving the way for the graphical user interface that we use today.

Remember, at the time, computers were instructed by feeding in punch cards or tape and the only output was beeping, printouts, or flashing lights. In this context, you can see how revolutionary real-time graphical object manipulation with a stylus would have been. It was basically magic. Computer pioneer Alan Kay once asked Ivan Sutherland: "How could you possibly have done the first interactive graphics program, the first nonprocedural programming language and the first object-oriented software system...all in one year?" Sutherland said, "I didn't know it was hard."

Interestingly, Ivan Sutherland was the mentor of Edwin Capmull. Capmull created the first computer-animated film for a class project in 1972 and went on to be the founder of Pixar. More on that in the next chapter.

Opening Computers up to Everyone: BASIC (1964)

In the 1960s, all computer coding must have looked like chaos to an outsider. That would all change in 1964, when two Dartmouth professors debuted a coding language designed to be easy enough for anyone to use. The new language was called the "Beginner's All-purpose Symbolic Instruction Code," or BASIC for short.

Before BASIC, computer programming, and subsequently computer use, was beyond complicated. Early mainframe computers were programmed as they were assembled, almost like an Ikea set, except you needed to know how to make the pieces talk to each other in code. In fact, only highly specialized scientists and mathematicians bothered learning how to write and use code, and hence were the only ones who used computers. Though coding was still a specialized task, BASIC made the process much easier.

John Kemeny and Thomas Kurtz, mathematics and computer science professors at Dartmouth College, and creators of BASIC, wanted computer coding to be more accessible.

The main idea was this: why not use everyday language that most people could intuitively understand? For example, HELLO and GOODBYE, rather than LOGON and LOGOFF. Even more important was that BASIC needed to work as a compiler. A compiler translates the written language humans understand to the 1s and 0s that the computer can understand. Previously, each user had to keypunch a program into cards by hand, and then wait for the results (which might take a day) as the computer translated the code line by line. Now, a user typed plain-English commands into the computer and BASIC converted them instantly. This was much faster and eliminated the laborious hand-coding of the time.

BILL GATES GETS HIS START WITH BASIC

One of the first to explore the full potential of BASIC was Bill Gates. Gates was born into a wealthy Seattle family. His father was a prominent lawyer, while his mother served on the board of directors for First Interstate BancSystem and United Way. The Gates family was competitive in nature, and the young boy liked to dominate in the strategic board game, Risk—skills that would come in handy later in his life.

During his childhood, Gates' parents began to have concerns about his behavior when he was around twelve. It wasn't that he was hanging out with the wrong crowd—actually the opposite. Gates was doing well in school, but he spent a lot of time on his own reading books.

His parents perceived him to be bored and withdrawn, but an event in 1968 would change his life, and history.

In 1968, the thirteen-year-old Gates encountered a computer while attending Lakeside High School. The school had a timesharing arrangement with General Electric (GE). This meant Lakeside paid GE to use the computer for a time. However, the mainframe computer remained at GE, while the school could view the output on a terminal screen. The practice was popular in the 1960s as computers were too large and expensive for most people to own. It was almost impossible for anyone who wasn't a scientist, an engineer, or at a university to get to see a computer in person, so this opportunity was a big deal for Lakeside High.

Gates just loved programming the GE machine in BASIC. Gates would later say that he was fascinated by the machine and how it would always execute software code perfectly. He was so keen that he was allowed to skip math class to go and code. His first piece of code was a computer-versus-human version of tic-tac-toe. Gates had found his calling.

Later that year, the school lost its funding for the timeshare, causing Gates and his (newly-acquired) coding friends to look for a new computer to play with. They found a PDP-10 belonging to Computer Center Corporation (CCC). At first CCC was happy to let them use the machine, until the four Lakeside students—Bill Gates, Paul Allen, Ric Weiland, and Kent Evans—were caught exploiting bugs in the computer's code to get free computer time. They were banned for the summer for their efforts. When the ban was over, the four students made a deal with CCC: "We'll fix the bugs in your code if you let us have more time to play with the computer." CCC agreed, and the arrangement was in place until the company went out of business in 1970.

As you would know, this isn't the end of Bill Gates' story. As we'll see in the next chapter, he and Paul Allen used BASIC to found Microsoft. The company laid the foundation for the PC revolution and Apple built on it.

While the workings of computers were being reimagined, the inner workings of psychiatry would be reconceived by a notorious chemical that broke out from the labs into the world during the 1960s: lysergic acid diethylamide (LSD).

FAR OUT: LSD LEAKS INTO CULTURE (1964)

No other chemical has impacted our understanding of the brain as much as LSD has. This chemical substance was at the center of a revolution that defined a decade.

Lysergic acid diethylamide made its original debut over two thousand years ago. In ancient Greek culture, it was a tradition to make an annual pilgrimage to a secret location to consume a drink made from fermented ergot (a fungus that grows on diseased kernels of rye), which produces a chemical similar to LSD. Some historians hypothesize that these experiences influenced ancient Greek thinkers.

As mentioned in chapter 7, LSD entered the modern world in the '40s. It was synthesized and discovered by Swiss chemist Albert Hoffman in 1943…by accident. The discovery occurred while Hoffman was trying to create a blood flow stimulant from the ergot fungus and touched his finger to his mouth. The next twelve hours were interesting, to say the least. He records:

"I had to struggle to speak intelligibly. I asked my laboratory assistant to escort me home. On the way, my condition began to assume threatening forms. Everything in my field of vision wavered and was distorted as if seen in a curved mirror. I also had the sensation of being unable to move from the spot. Nevertheless, my assistant later told me that we had traveled very rapidly."

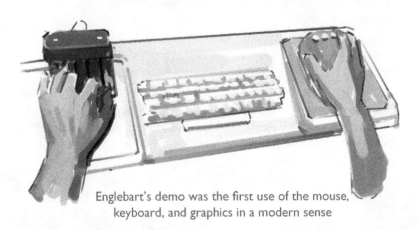

Englebart's demo was the first use of the mouse, keyboard, and graphics in a modern sense

At home, his condition became alarming: "My surroundings had now transformed themselves in more terrifying ways. Everything in the room spun around, and the familiar objects and pieces of furniture assumed grotesque, threatening forms. They were in continuous motion, animated, as if driven by an inner restlessness."

Hoffman believed his discovery could be a powerful tool for the scientific study of how the mind works. His parent company, Sandoz Pharmaceuticals, began to produce the drug.

At the psychiatric unit at St. George's Hospital, London, psychiatrist Humphry Osmond and his colleague John Smythies learned about Albert Hoffman's discovery. The pair conducted experiments to study the effects of mental illnesses, such as schizophrenia. This was done by giving LSD to healthy subjects and studying the effects. The fact that a chemical could induce schizophrenia-like symptoms flipped the medical field on its head. According to pharmacologist David Nichols, this shift in thinking could be said to be the start of neuroscience.

Osmond and Smythies looked at the chemical structure of lysergic acid diethylamide and realized that it was similar to the neurotransmitter serotonin. Could it be that the LSD was interfering with serotonin in the brain because their shapes were so similar? This was a light bulb moment that led them to conclude that psychiatric conditions are caused by chemical imbalances in the brain, an idea still widely believed today. In fact, antidepressant drugs such as Prozac act on the serotonin receptors in the brain—a direct consequence of early LSD studies. After this revelation, psychiatrists began to abandon the psychoanalytical approach in favor of new disease models based on brain chemistry.

David E. Nichols, a PhD student who studied the breakthroughs of early LSD studies, put things into context: "Up until that time, mainstream psychiatry had no idea that behavior might arise from neurochemical events in the brain. Rather, if parents had a schizophrenic child, the mother might be blamed for not being nurturing enough, or for doing something wrong in the parenting of the child. Parents all across the world and particularly women, who bore the brunt of childrearing, shouldered the guilt for a child with mental illness, including schizophrenia, believing that they had somehow failed as parents. It seems difficult to imagine such thinking today, but that was the reality of mainstream psychiatric theory back then."

In 1954, Aldous Huxley, a British writer, volunteered for one of the Osmond studies. He was so amazed by the effects that he began to encourage others to experience it. Huxley wrote a book called *The Doors of Perception*, exploring ideas resulting from the experiment. *The Doors of Perception* gained attention from a wide audience, including the CIA and the Soviets.

LSD Takes a Bad Turn

Military authorities from both the USSR and the US pored over *The Doors of Perception*. Both sides thought that the chemical Huxley had described could perhaps be used in warfare as a truth serum; they began testing shortly after. To their dismay, LSD's effects were so uncontrollable that testing proved useless. Subjects couldn't even concentrate enough to answer questions. This gave the US authorities another idea. Perhaps they could use the drug to make enemy leaders embarrass themselves in public. The plan was to slip it into the enemy's tea or coffee just before a large public event.

To test the idea, agents stealthily slipped LSD into each other's drinks at the office. This took a bad turn when agent Frank Olson died sometime after one of these experiments. The death was said to be related to lasting effects of the drug after an unpleasant experience. LSD proved to be powerful, but also highly dangerous. In 2017, the controversy around his death became the subject of the Netflix series *Wormwood*.

Around the same time, therapists in Hollywood began to use the chemical to help Hollywood actors and celebrities with their problems.

The Beginning of the End for LSD Research

Respected Harvard lecturer Timothy Leary first came into contact with LSD in 1962, and experiencing it, he made it his mission to use the drug to change the entire political, religious, and cultural structure of America. Other researchers were concerned by Leary's approach. They believed LSD should be used only for research under strict regulations.

Despite the protests of Leary's colleagues, Leary spread his ideas and young people across the country became intrigued by his experiences. In 1963, Leary was fired from Harvard. With nothing holding him back, he went from being a respected academic to a scruffy-haired preacher.

His message was to "Tune In, Turn On, and Drop Out." "Tune in" to the world around you, "turn on" to new levels of consciousness, and "drop out" of society's forced and outdated commitments. It was a message that resonated with the counterculture of the time.

Meanwhile, on the West Coast, Stanford student and soon-to-be famous writer Ken Kesey heard about all the fuss and decided to take part in CIA-sponsored LSD trials. Shortly after the trials, Ken applied to work on a psychiatric ward. During one of his shifts, he ingested LSD from the establishment's supplies. The experience inspired him to write the classic novel *One Flew Over the Cuckoo's Nest*.

Like Leary on the East Coast, Kesey decided to spread the word about LSD to America's West. He gathered a group of friends and drove around in a heavily-painted bus to distribute the chemical in the San Francisco Bay Area, and across the country.

Thanks largely to Ken Kesey and his friends, LSD soon exploded into American society on both coasts. The time was right: the leading edge of the baby boomer generation would turn 19 in 1964. They wanted their voices to be heard and they were questioning everything—they became the counterculture for the entire nation.

By this stage, Albert Hoffman, who had discovered the chemical, looked at the unfolding events with helplessness. He would refer to LSD as his "problem child." On October 6, 1966, LSD was made illegal in California by the state government—but this just made it more attractive. Bands like the Doors, the Beatles, the Rolling Stones and the Grateful Dead had started to spread the underground message of LSD and there was no putting it back into the lab.

Everything culminated in 1967 with the Summer of Love. 100,000 young people flocked to the Haight-Ashbury district in San Francisco and the media followed. These young, outlandish people were referred to as the hippies. Conservative Americans were horrified at what they were seeing. In 1968, the drug was finally made illegal in all fifty states. In 1969, LSD had one of its most infamous moments in the spotlight during the legendary Woodstock concert, when Wavy Gravy famously warned the crowd not to take the brown acid. By this time, the drug's darker side had been pulled into the light. There were scores of missing runaway kids in the San Francisco area, and many who abused the drug were left with lasting damage, or dead.

Albert Hoffman distanced himself from the chaos of the '60s and would live to the ripe old age of 102.

A NEW WAY OF DOING THINGS: THE MOTHER OF ALL DEMOS (1968)

Only a few blocks from the Haight-Ashbury area and the cradle of the counterculture, a very different kind of mind-blowing experience was occurring. On December 18, 1968, at the Graham Civic Auditorium, Douglas Engelbart (protégé of Lick from ARPA) would showcase a demonstration of the future of computers so far ahead of its time that it would earn the moniker "The Mother of All Demos."

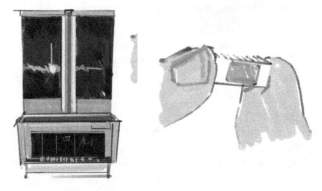

Typical 1969 computer VS a CPU

Remember Sketchpad by Ivan Sutherland? The drawing computer with the light-pen earlier in the chapter?

Well, Ivan's work and Bush's "Memex" machine inspired Engelbart to think differently about human-computer interaction. What if the computer could be a personal aid? What if it could be a planner, organizer, and communication device? And most of all, what if it were so easy to use that anyone could immediately figure it out without being taught?

The Mother of All Demos

There is a recording of The Mother of All Demos on grainy black-and-white film. It is fairly easy to find, and worth a watch. After Engelbart's opening words to the crowd, he states: "Imagine you're in your office as an intellectual. What if you had a display backed up by a computer that was alive, and all day, it was instantly responsive? How much value could you derive from that?"

Before Gates, Microsoft, or even Xerox, there was Engelbart. In this 1968 demo, Engelbart had just introduced the idea of the first graphical PC years before its time, and he was about to demonstrate how it worked.

As Engelbart begins the demo, he opens a blank document on a display and types a few words onto the screen. This part wasn't actually revolutionary, albeit a recent invention at the time. The next thing certainly is. Once the words are on the screen, Douglas uses a mouse to copy and paste some of the words. After this, he goes on to demonstrate how the page can be arranged and formatted. This was the very first use of a computer mouse, and it was used in (what can be considered) the very first word processing program.

Engelbart asks the computer to save what he's written into a file. Next, Engelbart "right clicks" on the file to open up the equivalent command of "viewing file properties."

Following this was a typed shopping list. Engelbart shows that he can organize items by categories and click to open a particular category. He next goes over various ways of manipulating the shopping list.

Engelbart then performed something which he called "jump on a link" to view more details about the items on his shopping list. This was essentially hypertext; much like the hyperlinks we visited before, they take you from one page to another on the internet, but by using text.

Later in the demo, Engelbart states that his team had been using the system for a few months and had six computers connected together. In other words, an early, closed-loop, intranet (as opposed to the open *inter*-net of ARPANET days). He even goes on to demonstrate a form of email between computers. ARPANET would still be a year away at this stage.

For 1968, this was mind-blowing. Any one part of this demonstration would have left an audience speechless, but Engelbart kept piling innovation onto innovation. This was truly The Mother of All Demos.

Engelbart had developed the mouse, hypertext, onscreen windows, and many other features of modern software. Again, Vannever Bush is influential here. Engelbart credits Bush with awakening him to the potential of computers to manage information, not just crunch numbers. As incredible as this was, the significance of connected graphical computers wouldn't be realized until the mid-1990s.

Although graphical computers were off to a slow start, computers were just about to get a whole lot faster and smaller.

"A WHOLE COMPUTER ON A CHIP!": INTEL (1968)

In the previous chapter, we discussed the formation of Fairchild Semiconductor (founded by the Treacherous Eight) and the invention of the integrated circuit in Silicon Valley. With this technology, Fairchild was way ahead of then-tech titans IBM and Motorola.

Each passing year brought new companies to Silicon Valley, and with them new jobs, though Fairchild Semiconductor was still the Valley's star attraction. Workers loved the unique corporate culture fostered by their once-reluctant boss Robert Noyce. It was a radical departure from other businesses. Noyce wanted to break down the distinctions between management and workforce. In his domain, everyone was a part of the team. People would happily work ten- to twelve-hour days, just because they enjoyed it.

At first the culture was like a college dormitory for competitive geniuses, and staff were making big money competing with each other. Soon, Fairchild was on a meteoric rise. They were like the Amazon, Apple, or Google of their day, both on the stock market and in the technology space. But in time, it became clear that there was too much talent there for one company, and by 1962 Fairchild Semiconductor was beginning to unravel.

Half of the founding team and numerous researchers and engineers left the company to start new ventures. Most of them either worked for or became direct competitors in the integrated-circuit market. As ex-employees of Fairchild moved on and spread what they had learned, Silicon Valley exploded. The effect was so prolific that these ex-employees were nicknamed "the Fairchildren." By the 1980s, 100 different companies had sprung from the talent pool of Fairchild. But even as the dam burst, Fairchild was still weathering the storm. In 1964 alone, the company had shipped 100,000 devices to NASA, but Noyce was eying a bigger target—the American consumer.

As the innovations of Silicon Valley piled up, people with money wanted to invest in the hottest and fastest-moving technology in human history. This was the idea of venture capitalism that we touched on before.

As more and more people left Fairchild, companies like Texas Instruments and Motorola started catching up. Noyce realized he needed to start something new. In 1968, he and fellow employee Gordon Moore began a project to make memory chip devices, to compete with the magnetic hard drive. Noyce and Moore's status in the industry, coupled with the rise of venture capitalists, resulted in $2.5 million being raised in under two days for their project, resulting in the formation of a new company.

Noyce and Moore picked the best and brightest minds to join the new company. Among them was Andy Grove, a Hungarian-born chemical engineer who had joined Fairchild's R&D division in 1963. His new role was Director of Operations. Former employees would say that hiring Andy was the best decision Noyce ever made. He was a driven man who wanted everything to be done well and done on time. Neither Noyce nor Moore were stern enough to crack the whip.

Together, they called their new company Intel, an abbreviation of "integrated electronics," which also happened to conjure the word "intelligence." It didn't garner a lot of attention from the media, but within the Valley there was a great deal of excitement.

In the spring of 1969, as Intel engineers continued to tinker with the design of their memory chip, a simple request came into the Intel building, altering the course of computer history.

The Computer's Brain (CPU) Is Built

A Japanese firm, Busicom, contracted Intel to design twelve specialized microchips for its new calculator. Unfortunately, what Busicom wanted was going to be way too complicated and expensive for Intel to produce. A young engineer named Ted Hoff raised his concerns. He thought that Intel might have bitten off more than it could chew.

Ted Hoff was a smart guy. While at Stanford in the early 1960s, he had spent four years conducting research on pattern recognition (what would later be called neural networks) and integrated circuits. Hoff had joined Intel in 1968 as employee number twelve. When Hoff voiced his concerns to Noyce, Noyce replied that—if Hoff could think of one—he should pursue a simpler chip design than what Busicom had requested. Noyce always encouraged the people in his labs to run with their ideas and see where they went.

Hoff envisioned a single chip that could be programmed for a specific application. In this instance, it would be to function as a calculator, but it could also be programmed to do *anything*. If the architecture could be constructed simply, then the memory, calculating, and processing functions of a computer could be combined into one integrated circuit.

Their efforts led to the first microprocessor: essentially a whole general-purpose computer on a chip.

The result was the Intel 4004 central processing unit (CPU). With more than 2,000 transistors, the device was advertised as a "computer on a chip." In this moment, the digital revolution had officially begun. The CPU is the culmination of all previous computing innovation. It is seen as one of the most important inventions of the last hundred years. The CPU was to become the driving force behind the brain of all of any computational aid we use today.

At just an eighth of an inch wide (3.2 mm) and one-sixth of an inch long (4.2 mm), the chip Hoff and the team unveiled had as much power as one of the first electronic computers, the pre-transistor ENIAC. The ENIAC of the 1940s used 18,000 unreliable vacuum tubes and was so large it filled an entire room. Though computers had shrunk since the days of the ENIAC, the 4004 was still a giant leap ahead of what was available at the time. Hoff's chip took a 1969-era computer that was the size of a refrigerator and shrunk it down to fit on a fingertip!

Amazingly, some still had reservations. Japanese clients were concerned about multi-tasking. For example, they were worried that the chip wouldn't be able to show a loading light and perform calculations at the same time. Hoff showed them that they were wrong. All they needed to do was write a program that would cause the lights to blink and do the processing, at the same time (or switch between the two so fast that the human brain would never know). What the chip did was scan the keyboard for input, then output anything it found onto the display of the calculator.

This was done using something called micro instructions, or microprograms. The microprograms were hand-coded into a read-only memory chip (ROM)—something Intel had also just invented!

Intel would call the ROM the "personality" of the system, as it is the part that tells the CPU how to behave. To change how a computer behaved, you simply had to re-program the ROM. Traditionally, you'd have to undergo laborious rewiring of the individual transistors or change out large circuit boards. It's hard to imagine doing that today.

The fact that the CPU could switch between calculating on the Japanese calculator, and blinking a light, is a key concept. A modern computer's CPU works by taking small tasks and switching between them hundreds of thousands of times per millisecond. It does it so fast, it gives us the illusion that it's doing multiple things at once.

When the 4004 microprocessor (CPU) was unveiled, people couldn't believe it.

Ted Hoff would recall one such interaction at a computing conference in Las Vegas, 1971: "One customer who came in was adamant that we had such nerve to claim we had a computer on a chip. One of our engineers handed him the datasheet and he took it and exclaimed 'Oh God, it really is a computer!' It was something that people did not believe was possible at the time."

A slightly later version of the chip, the Intel 8080, would be the heart of the first widely-used "microcomputer" by MITS. This Intel-powered computer started both Apple and Microsoft (Microsoft stood for "Microcomputer Software"). More on this in the next chapter. Today, the microprocessor market is a multi-billion-dollar industry. Intel is currently the largest CPU manufacturer on the planet and still dominates the desktop and laptop markets.

As for Ted Hoff, he left Intel in the early 1980s to accept a position with Atari as Vice President for Technology. Robert Noyce left Intel in the late '70s. Today, Noyce is eighty-nine years old and thoroughly enjoys fishing.

The 1960s: what a decade! The state of computing had changed drastically by the end of ten short years. Culturally, the baby boomers had changed society, and in technology, new ideas that would impact humanity forever had come to fruition.

In the next chapter, the pace of innovation seemingly doubles as technology leaks into mainstream culture.

CHAPTER 10

The New Age
1970–1979

The 1960s shook the world with its radical new ideas, but the 1970s was an abrupt change in the overall mood within the social fabric of the Western world. The post-war economic boom was at an end, oil prices were up due to crises in the Middle East, the Vietnam War continued to escalate, and the Watergate scandal led to the impeachment of Richard M. Nixon. Many took to the streets to further civil rights movements such as women's rights, race equality, gay rights, and others. There was also voiced disapproval of the US government, while the New Age movement took hold; this movement led to (what became known as) the "me" decade.

The second half of the '70s lightened up a bit, as the economy improved slightly, and a new form of music called disco began to emerge. It started off as extended remixes of blues and R&B records that people could dance to without interruption. Rock was also seeing a change and, by the end of the decade, glam rock and power ballads had come of age.

In the '70s, email made its debut, along with the first word processor, the C programming language, and the floppy disk. Gene Cernan would be the last man to set foot on the moon in 1972. In 1975, an unknown young director burst onto the international film scene with his film *Jaws*. The director, Stephen Spielberg, would go onto become one of the greatest minds in cinema. In 1976, the most successful supercomputer, the Cray-1, was commissioned. It ran at a blistering 80 megahertz (twenty-five times slower than a single processor in a modern smartphone) but was a breakthrough for its day. The '70s also gave us everyday inventions like the modern soda-can design, the Post-it note, and the Rubik's Cube. But perhaps the greatest achievement of the decade came in 1977, in the form of NASA's Voyager missions. Also in 1977, all the planets happened to align—an event that happens once every 176 years. This cosmic event provided a unique opportunity to study all the planets.

Voyager 2 was the first spacecraft to explore Uranus and Neptune. Studying Uranus, the probe discovered ten new moons and also found that the planet was tipped on its side. Six new moons were discovered orbiting around Neptune, as well as a storm the size of Earth flying around the planet every eighteen hours. Almost all the information we know about those planets came from Voyager 2. Other discoveries by the spacecraft included the nine volcanoes on Jupiter's moon, Io. These powerhouse volcanoes produce plumes of smoke that rise 300 km into the air, at velocities of 1 kilometer every second! The Voyager 2 craft also managed to get so close to Saturn, it could detect that it was less dense than water. This meant that, if you had an ocean big enough, Saturn would float in it.

The Voyager craft are still in operation today and run on a plutonium isotope. The probes are transferring data back to earth at speeds slower than a dial-up modem, traveling an incredible one billion miles every three years.

Aside from space travel, this decade was responsible for ushering in our modern way of living. The '70s was the start of a different kind of new age.

Genetic Engineering (1972)

James Watson, Francis Crick, and Rosalind Franklin (the forgotten X-ray wizard) discovered the double helix structure in 1953. Strangely, it would take around twenty years for this information to have a practical use. In 1972, Herbert Boyer and Stanley Cohen created the first genetically-modified organism, a bacterium that was resistant to kanamycin (an antibiotic). Shortly after this, Rudolf Jaenisch created the first genetically-modified animal.

HOME VIDEO GAMING FINALLY ARRIVES (1972)

Today, almost all of us either own a game console or have played on one. From Nintendo to the Xbox or PlayStation, there have been countless hours of entertainment provided by home game consoles. Higinbotham's *Tennis for Two* in chapter 8 and MIT's *Spacewar!* in chapter 9 were both standalone attempts at early computer gaming, but who was the fine human who invented home gaming?

To find out, we have to travel back to 1966.

The TV "Game Box"

In 1966, a man named Ralph Baer was the head of equipment design at Sanders Associates, a military company. One day, while waiting for a bus, Ralph was thinking—as there wasn't much else to do. During this wait at the bus station, an idea came to him: the simple concept of a "game box" that could turn a regular TV into a gaming device. Inspired, he wrote up a short proposal for such a device the very next day.

What was he to do with such an idea? A gaming device had nothing to do with the military, so his bosses would be less than impressed if he showed them. Knowing this, Ralph stole an assistant by the name of Bob Tremblay from the Sanders company and together, they got to work in an empty room.

Soon, they had a prototype creatively called "TV Game #1" and honestly, it wasn't very good. The "game" consisted of a single vertical line that could be moved across a TV screen. Despite not being much to look at, TV Game #1 was demonstrated to the director of research at Sanders, Herbert Campman. After raising an eyebrow and probably exclaiming, "Sure, what the heck," Campman agreed to allocate $2,500 ($19,332 today) for funding.

In the next year, Baer and a very small team worked on improving the console. Eventually, they came up with a console which was capable of playing a shooting game with a plastic light gun. Ralph showed the new game to Herbert Campman, who was now enthusiastic.

Campman increased funding and pushed the idea to senior management, who largely just grunted at the idea. Still, the Sanders CEO thought a TV game console was a product that could be marketed and sold. By 1969, the small team was approaching a definitive version that had almost thirty games—they just needed someone to sell it to. The television manufacturers would be a logical target. After some failed negotiations with RCA, in 1971, Baer's team settled with the Magnavox company, a media hardware powerhouse of the day. The launch date was set for September of 1972, and the game console was to be named the "Magnavox Odyssey." The console sold for $99.99 ($600 today) by itself, or $50 ($353 today) when bought with any Magnavox television. This was the first home gaming console.

Since the term "video game" didn't exist at the time, during the advertising campaign, the system was referred to as a "closed-circuit electronic playground." Catchy.

So, what did the public think about such a device? The only form of visual home entertainment for most at the time was black-and-white TV. Suddenly, they would be able to instantly interact with a TV and play games? It must have seemed like the future had come all at once! But…the sales didn't really reflect that.

Due to the high price, and confusion in the advertising (which implied that the consoles only worked with Magnavox TV sets), sales remained modest until the console's cancellation in 1975. Worldwide sales only totaled around 350,000 units. Regardless, the Odyssey ushered in the modern era of gaming and directly inspired Atari.

Hey, Don't Steal My *Pong*!: Atari

The story of Odyssey and the Magnavox system didn't stop with this home game console. Around the time of the Odyssey's release, Nolan Bushnell was chief engineer of an arcade game manufacturer. Bushnell had previously come up with an arcade game called *Computer Space* (essentially an arcade version of the 1962 MIT rocket shooter game, *Spacewar!,* from the previous chapter). *Computer Space* was also the first ever commercial arcade game. The cabinet-like structure housing the game used custom hardware specifically built to run the game, as it couldn't run on the cheaper minicomputers. The cabinet was put in a few local bars but, unfortunately for Bushnell, the game wasn't really a success due to its complexity (or maybe people were too drunk to play properly). Regardless of the reason for failure, Nolan needed another idea.

Magnavox Odyssey

One day in May of 1972, he came across a demo of the Magnavox Odyssey at a dealership. From the moment he saw a table tennis game being played on a TV screen, he knew that things were about to change. He immediately quit his company and started a new one called Atari, hiring a man named Allan Alcorn in the process.

This is where Bushnell becomes a sneak. Without telling Alcorn that the Odyssey demo belonged to Magnavox, Bushnell asked Alcorn to devise a similar ping-pong-type, coin-operated arcade game as part of his "company training." What Bushnell really wanted was a rip-off ping-pong game for which Alcorn did all the work.

Alcorn was a bit of a genius and decided that the original version of the game was too boring. To make it more exciting, he added a few features of his own. Instead of the ball just going back and forth in a straight line at a constant speed, he thought the direction of the ball should change depending on how the player hit it. In addition, he thought that the ball should speed up the longer you played, and only reset to its original speed once you missed. This game would be known as *Pong*.

The *Pong* arcade machine was a simple setup—a black-and-white TV purchased from a department store, some custom hardware that ran the game, and a laundromat coin mechanism with a milk carton to catch the coins.

In August of 1972, the game was put in a local tavern to see how people would react to it. It seemed like Bushnell and Alcorn wouldn't get much of a chance to see, because a few days after installation, the game began to malfunction. Alcorn got the call to fix the machine and was on his way. On the drive, Alcorn must have had some questions going through his mind: What went wrong? Maybe it was bad coding? Or maybe the hardware in the cabinet wasn't put together properly?

When Alcorn arrived at the tavern, he was stunned to see a line of people waiting outside waiting for the bar to re-open. They were all waiting to play *Pong*!

As for the problem? The game was so popular that when the coin collector became full, eager patrons stuffed in so many coins that the mechanism shorted out and shut down. The Atari pair knew that they had a smash hit on their hands.

Pong quickly became the latest trend for the kids of America, and the arcade became the hottest place to hang out. Bushnell would later recall that many kids met their future partners over a game of *Pong*.

Pong, and to a lesser extent the Magnavox Odyssey, truly kicked off the modern era of gaming. In 1974, Magnavox would sue Atari for stealing the idea of *Pong*. Atari agreed to pay a settlement.

The Hand That Started Pixar (1972)

We're used to seeing 3D animation everywhere today, but there's a curious story behind its inception. Its birth has to do with the ingenuity of one man, Edwin Catmull. Catmull was born as a post-war kid in 1945. His idols were Einstein and Walt Disney. Movies like *Pinocchio* and *Peter Pan* captured his imagination. He wanted to be an animator, but didn't know how, so he took up computer science at the University of Utah in a nod to his idol, Einstein. During his graduate studies, he was a student of Ivan Sutherland (the man who invented Sketchpad, the first computer drawing program, discussed in the last chapter). Catmull saw that animation on a computer might be the perfect way to marry his love of science and art.

From that point, Catmull's main goal and ambition was to make a computer-animated movie. His first opportunity came when the class was given a task of creating a digital 3D object. When Catmull observed the sad state of computer animation, he realized that the software was inadequate to allow for animations suitable for the film

Impression of Ed's 3D animated hand

industry. In a stroke of brilliance, the young man decided to make his own animation software, creating texture mapping, spatial anti-aliasing, and z-buffering, thus singlehandedly creating the fundamentals of computer graphics along the way.

For his project, Catmull decided to digitize his left hand by creating a plaster-of-Paris mold. Unfortunately for Catmull, the mold was on so tight that when he removed his hand, he also removed all the hair on the back of his hand. After the mold removal was (painfully) done, Catmull traced 350 triangles onto the surface of the hand. The edges of these triangles would represent coordinates, which had to be painstakingly entered into a teletype keyboard for the computer to process.

The computers of the time weren't capable of calculating an entire frame of footage at once, so the output on the screen was captured with a long-exposure camera until the entire frame was processed. When this was done, a snapshot would be taken. A lot of work for just one frame.

The final video, simply known as "A Computer Animated Hand," was eventually picked up by a Hollywood producer and incorporated in the 1976 movie *Futureworld*, the science fiction sequel to the film *Westworld* and the first film to use 3D computer graphics.

Catmull later went on to form his own computer graphics lab and worked with the great George Lucas to bring computer graphics to entertainment. Catmull's graphics lab (Graphics Group) later spun off as one-third of the computer division at Lucasfilm and was renamed "Pixar." The company was heavily invested in by Apple Computer co-founder Steve Jobs in 1986.

It's strange to think that the roots of Pixar can be traced back to a hairless left hand. Computer graphics are ubiquitous today, but even more so is the mobile phone, another invention of the 1970s.

THE FIRST MOBILE PHONE (1973)

Phones have altered our expectations of what's possible. Now we expect to have all things with us at all times; the entire world is at our fingertips, instantly. We can read, watch, listen, consume, and sometimes create, from anywhere at any time.

What I've just described is the smartphone revolution, and it's only a decade old. But, before we had the world in our pockets, mobile phones did a lot less. In the '90s and '00s they had a basic organizer, a modest camera, and very basic Internet if you were lucky, but go back to the '80s and mobile phones just did one thing: make phone calls.

What was the first one of these primitive phones that just made phone calls? What was the original phone that would become the grandfather to all the billions of phones in our pockets? The person who invented it might just be worth checking out.

To start the story of the first mobile phone, we have to go back to the 1940s. Technically, at this time, mobile phones did exist, but only in cars. And if you had one, you were most likely an important person. Around this time, putting a phone in a vehicle was the only way to make it mobile. This is because the phone took so much power to run that only car batteries could meet the supply.

Another drawback was that, for a given area (New York for example), only twelve phone-line channels existed, so most of the time you'd have to wait to use a network. Just to connect to a mobile call in a car could take thirty minutes. That's just the way it was and the way it was always going to be. Well, until it wasn't.

In 1968, the Federal Communications Commission asked AT&T to fix this issue. AT&T then came up with a cellular radio architecture.

Its aim was to break up the large areas of coverage into smaller ones so that multiple people could use their phones in their cars at the same time. You see, at the time, regular mobile systems used a central high-power transmitter to serve an entire city, so only one telephone call could be handled on a radio voice path at one time. The new system would use a number of low-power transmitters arranged in "cells" (that's where the name "cell phone" comes from). The calls would be automatically switched from one cell to another as a driver moved.

Motorola also had a car phone division and didn't want AT&T to have a monopoly on products that could take advantage of this new system. Motorola feared that this could be the end of their mobile business if they didn't do anything. So, they began to develop a phone that took advantage of the new cell system. The device was to be so advanced that it would leave AT&T in the dust. In 1972, Motorola asked an employee by the name of Martin Cooper to spearhead the project.

DynaTAC 8000x

One day, while thinking about how to create a mobile phone, Cooper recalled a comic strip called *Dick Tracy*. In the comic, Dick has a mobile wristwatch he uses to help fight crime.

This gave Cooper an idea and he began to think about the problem differently: instead of assigning a mobile number to a *place*, why not assign it to a *person*? Then you could be connected wherever you are, to whomever you want. This was a revolutionary idea—but could such a device even be made? Cooper was banking on a brand-new technology. The microprocessor, freshly invented by Intel in 1971, was the key. Its tiny form meant it could do the same thing as many components, using much less space and power consumption.

By March of 1973, Cooper and his team had a working prototype called the DynaTAC. On April 3, 1973, Cooper introduced the DynaTAC phone at a press conference in New York City.

The First Phone Call Was a Troll

For his first test call before the press conference, Cooper decided to call an engineer named Joel Engel. Engel was the head of AT&T's mobile phone project, the arch nemesis of Motorola. So, Cooper picked up his new phone and called Engel to tell him that he'd beaten AT&T to the mobile phone. That's right, the very first mobile phone call was a troll. Cooper explains:

"In that first call we didn't know it was going to be historic in any way at all. We were only worried about one thing: is the phone going to work when we press the button? Fortunately, it did."

In 1983, after years of development, Motorola introduced the first portable cell phone to consumers—the DynaTAC 8000x. It weighed almost one kilogram. Additionally, a full charge took roughly ten hours, and offered only thirty minutes of talk time. But that was okay, because your arm would grow tired if you held it up to your ear any longer than that. To top it all off, its price was almost $10,000 (adjusted for inflation).

The phone eventually became a success, kicking off the mobile phone revolution. When Cooper made the first mobile phone call back in 1973, there was only one real cell phone. Now there are more mobile phones than people. While Cooper was making the phone call that changed the world, there was another event that would alter history: the first *graphical* consumer PC.

Snatching Defeat from the Jaws of Victory: Xerox Alto (1973)

Imagine having one of the greatest inventions of the twentieth century in your hands and giving it away because you didn't understand what you had. Xerox did just that with the Xerox Alto.

The Xerox Alto was an experimental computer from 1973, created at Xerox's Palo Alto Research Centre (PARC). The PARC team consisted of Robert Taylor—a researcher from ARPA of the last chapter—and other researchers who were inspired by Douglas Engelbart (Mother of All Demos) and Joseph "Lick" Licklider.

Xerox Alto was a modern desktop computer in every sense

The Alto was way ahead of its time. It was the first modern desktop PC to resemble what we recognize today. The central idea of the PC was based on Lick's 1960s paper "Man-Computer Symbiosis." It described how a computer could be like a friendly colleague that could help when problems got too hard to think through.

The Alto had a mouse, windows, icons, menus, graphics, file managers that could copy, paste, move, and delete files, and even a local network that connected computers together. This is commonly referred to as a graphical user interface (GUI). The idea was to mimic the office desk, but on a screen—a paperless office of the future. Certainly, a new way of thinking for 1973.

What the Xerox Alto represented was the very first graphical user interface in desktop form.

Although shown once in the Mother of All Demos, pointing at and actually clicking on a graphical object was a very foreign idea in mainstream computing. Before GUIs, to do anything on a computer, you needed to type commands out in lines of text. If you mistyped something, too bad. The computer would just spit out an error.

Thousands of Xerox Altos were built at the PARC, but never sold to the public; they were only used heavily at the other Xerox offices and at a few universities. Unfortunately, the Xerox top managers didn't understand what they had. They thought the Alto was an amusing toy, but nothing more. The managers didn't have a vision of what the computer of the future could become, but a man named Steve Jobs did, and Xerox handed the completed vision straight to him. Here's how it went down:

Despite protest from some employees, Xerox managers invited Jobs to look at the Alto in 1979. The meeting was arranged as an exchange: Jobs would help the Xerox team figure out how to make computers more cheaply (since Apple II's were flying off the shelves at the time), while Xerox would allow Jobs to take a look at some of their cutting-edge research at the PARC center.

During the meeting, Jobs was amazed. He knew that this was the future—everything that he imagined a computer to be was sitting right in front of him. Jobs didn't waste the opportunity and heavily borrowed these ideas of a graphical user interface featuring onscreen windows and a mouse into his next product: the Apple Lisa, which would lead to the original 1985 Macintosh. We'll revisit this story in the next chapter.

PERSONAL COMPUTING IS BORN (1975)

It goes without saying that the personal computer revolutionized businesses and homes around the globe. But where did this whole revolution start? As you'll soon find out, the first PC was the impetus for both the Microsoft and Apple companies.

In 1974, Ed Roberts, an engineer, was trying to stop his calculator business (MITS) from going under. At a meeting with his bank lenders, and in a last-ditch attempt, Roberts said that he might be able to save the company by doing something that hadn't been done before—creating a personal computer kit for hobbyists. This was a task thought impossible at the time. It was only made possible by the microprocessor (CPU), invented by Ted Hoff and company at Intel just three years earlier. As we saw, the microprocessor allowed for an entire computer to be put onto one chip, a huge feat. Intel had released the 8080 chip in 1972, a more powerful version of their initial product. Roberts saw an opportunity and used the silicon in his new computer, the MITS Altair 8800. It would be the world's first microcomputer kit.

Popular Electronics Magazine:
The Spark that Ignited It All

MITS advertised the microcomputer on the front cover of *Popular Electronics* magazine for $400 ($1800 today). For *Popular Electronics*, advertising a computer in a hobbyist magazine was a bit of a risk, as it wasn't known if it would even be a good fit. It proved to be exactly the opposite.

Roberts told the bank that MITS expected a few hundred orders at the most but ended up being swamped with a couple of thousand within the first few months. The demand was so high that enthusiastic hobbyists parked in the front of the MITS warehouse to get their hands on the computers.

Thinking about it, the demand was only natural. In the early '70s, computers were still a big deal. You usually needed a clearance to see one, let alone use one. This made it almost impossible to get to use one in person without timesharing (logging on to a terminal connected to a large mainframe computer). Even when you got there, you'd be frustrated because computers seemed to be getting more difficult to use as time went on. In the mid-1970s, computer technology began to evolve in capabilities beyond just calculations. It seemed that computers could now do a lot more than just math.

Globally, young tech minds were entranced by the concept of a computer. One of those young minds was the rebellious computer hacker Bill Gates, now in university.

Paul Allen, a high school friend of Bill Gates, saw the magazine cover with the 8800 sitting proudly on it. Allen snatched it off the shelf and ran into Gates' college dorm room. Upon seeing the device, Gates knew that this was it. A consumer personal computer had arrived. This event would change Gates' life, and the rest of ours too. In an instant, all the possibilities were now within reach.

Well...almost. You see, the Altair 8800 was very primitive. Early users of the computer would later say there was no place to connect a keyboard or monitor; it was just a bunch of switches and lights. In addition to being primitive, the Altair 8800 was difficult to use. To load an instruction, you had to flip switches. For example, to figure out 2 plus 2, each "2" required eight different switches to input and a ninth switch to load them all. The "add" instruction required eight more switches, and the answer "4" was displayed by the third light from the left turning on (I told you it was difficult). In addition to this, there was no memory. This meant you couldn't turn the machine off after you coded something because the information would be lost. We have to be thankful for our ease of use today.

Microsoft: Gates Turns His Hobby into a Company (1975)

When Paul Allan burst into Bill Gates' dorm room and showed him the MITS Altair 8800, Gates also realized that the computer revolution was about to happen without him. He called up the MITS company to tell them that he and Allen had written software for the machine and that the code was almost completed. The only problem was, they hadn't even started writing it.

The nineteen-year-old Gates saw that he could get an early foothold on the new PC industry. He quit university and went to New Mexico, where he lived at a hotel across the road from the MITS factory. From this hotel, he and Paul got to work writing software for the 8800. They chose to make a version of BASIC for the Altair so that it could be smart enough to perform functions other than flashing lights. The thing was, they didn't have a MITS Altair machine to do the testing on—only a book from Intel on the chipset that was used in the Altair 8800 computer. If they made one mistake, their software wouldn't work. Undeterred by the challenge, the pair wrote the software in a couple of months and punched it into paper tape. This piece of paper with holes in it was the very first Microsoft software—you'll recall that "Microsoft" (fittingly) stood for "microcomputer software."

Paul Allan then flew to the MITS factory and ran the software on an Altair 8800, with MITS CEO, Ed Roberts, as a witness. Allan typed in a program to print "2 + 2," and the computer spit out "4." He tried some squares and square roots, all successfully. The software worked. In the end it was Gates' business mindset, risk-taking attitude, and confidence in skills that allowed him and Allan to deliver something they didn't yet have, but knew the world needed.

Gates and Allan's software now allowed the Altair 8800 to do much more. By the end of the year, people had attached screens to the Altair, and written games, accounting software, and word processors on top of the early Microsoft BASIC software. This was where the Microsoft company originated.

A true revolution had begun, and at the center of it was Microsoft. And, as we know, they didn't do too badly after that.

APPLE COMPUTER (1977)

The other side of the PC coin was Apple. Today, they're one of the most valuable companies on earth, but how did it all begin? You know what they say—great things happen when you're in the right place at the right time.

The year was 1977, and the timing was perfect for an explosion of innovative technology. A remnant of the '60s counterculture remained in a revolution separate from mainstream America. In tandem with this was the development of the CPU from Intel. During this time, a sixteen-year-old Steve Jobs met a twenty-one-year-old Steve Wozniak. They were run-of-the-mill university students with a love of technology, pranks, and the Beatles. Neither of the friends had any idea that the two of them would soon change history.

At this time, the hysteria and fallout from the Altair 8800 on the cover of the *Popular Electronics* magazine caused computer clubs to pop up around the US. One of these clubs was the Homebrew Computer Club in San Francisco. The aim of the meetings at this particular club was to find a real "human" use for the new computer.

According to Jim Warren, founder of the West Coast Computer Faire, the general belief system of the Homebrew Computer Club was influenced by the counterculture prevalent around the area. For the club, this meant that helping each other and sharing everything was the norm. It was a team effort to figure out what the new computer kit could do. Members of the club who had purchased Microsoft software began to copy and share Bill Gates' paper tape software (by punching holes with a teletype) without his permission. They didn't see anything wrong with that, but Gates was furious!

This open behavior was at odds with what Bill Gates believed. He flew down to the Homebrew Computer Club with a presentation, begging people to stop copying the software. The club members didn't care much for Gates' protests. They were just excited to see what new functionality could be brought to the Altair machine with each inquisitive piece of code.

Wozniak (Woz) later stated that it was the very first meeting at the Homebrew Computer Club that inspired him to work on his own computer, which he brought to each weekly meeting. Woz was always a tinkerer and was even expelled for sending prank messages on his university's computer. When he was in college, in the early '70s, he would read booklets and design his own computers on paper, eagerly waiting for the chance to be able to afford a CPU chip of his own. He recalls:

> *"...and I took this book home that described the PDP-8 computer and it was just like a Bible to me. You might fall in love with playing a musical instrument; I fell in love with these little descriptions of computers on their inside. I could work out some problems on paper and solve them and see how it's done, and I could come up with my own solutions and feel good about it."*

When Woz told his dad (while in the first year of college) that one day he would have a desktop computer, his dad replied, "It costs as much as a house, son." Woz argued that he'd "just live in an apartment then."

By 1975, Wozniak was already building his own computers and completed his first prototype in June of that year. He would "borrow" silicon chips from his work at Hewlett-Packard, where he designed calculators.

The prototype computer Wozniak brought to the Homebrew Club each week could display letters and run a few simple programs. As simple as this seems today, this was the first time in history that any characters were generated on the screen of a home computer. Each week, his computer would have new improvements. Woz's main motivation for building the computer as efficiently as possible was to impress his buddies at the club.

Wozniak was too shy to put up his hand during the club sessions, so he would just set up his computer at the session's conclusion for people to check out. People would crowd around his creations and ask him a stream of never-ending questions. Everyone wanted one!

Steve Jobs—friend of Woz, and an Atari intern at the time—also attended the club meetings and noticed the demand. He said to Woz that the two of them should really sell these things.

Wozniak had proven that he could singlehandedly design a circuit board—all the required hardware and software that was needed for a personal computer—so the two of them could go ahead and do it. It just might work...

The First Apple Computer

Wozniak tried to sell his computer design to HP, but they rejected it five times. In the end, it would be their loss.

After Jobs suggested they sell the computer as an assembled circuit board, the pair decided to give it a go. Woz sold his HP calculator and Jobs his Volkswagen van to get the project off the ground. Together they raised $1,300 ($5,700 today) and, with the help of some friends, made their first personal home computer, initially working in Jobs' bedroom and then in his garage. This item was the maiden Apple product, the Apple I, which sold for around $600 ($2,640 today). The design used far fewer chips than any comparable machine and Wozniak earned a reputation as a renaissance artist of computer design. The Apple I was a single circuit board with no included keyboard or monitor.

In April of 1976, Jobs and Wozniak founded Apple Computer—this was the official beginning of Apple. In the same year, the company managed to sell fifty Apple I's. To Jobs, this proved it wasn't just hobbyists *within* the club that wanted to own a PC, but those without also. But what about the people who weren't hobbyists at all? What about the rest of us? Jobs explains:

"It was very clear to me that while there were a bunch of hardware hobbyists that could assemble their own computers—or at least take our board, and add transformers for the power supply, as well as a case and keyboard—for every one of those, there were a thousand people that couldn't do that, but wanted to mess around with programming. So, my dream for the Apple was to sell the first, real packaged computer."

I think a lot of other hobbyists would have missed this piece of insight. Understanding the increased potential of selling a complete PC may not have come easily to someone who was just a tinkerer—realizing this would take an individual with a natural business mindset.

Jobs' vision of a ready-made PC was actually impossible at the time. It would take way too many chips, making it too expensive and too complicated. As it turns out, Woz didn't know it was impossible.

Woz was a wizard in computer design. The fact that the Apple I had color was actually because of a trick he had learned at Atari for their early arcade games (Woz made the famous game *Breakout* there). Jobs contributed his design knowledge, making the computer attractive to look at, while insisting that the computer should be easy to use. Even looking back today, the sleek, tilted design must have been a refreshing change from the bland boxes that were available.

The Apple II

Steve Jobs managed to get some investment money together for manufacturing. The result was the Apple II, which sold for $1,300 (or $5,242 today). It was a fairly capable, ready-made home computer that only required the user to plug in three things: a keyboard, a monitor, and a power supply cable. The whole thing would be powered by Microsoft BASIC.

By 1977, the PC was a far cry from the hobbyist box with a bunch of lights and switches that featured just two years earlier—it was now a complete product ready for the consumer, without the messing around. It was the *real* genesis of the home computing era.

The new company showcased the Apple II at the West Coast Computer Faire in 1977. They stole the show outright with the Apple II's color display and built-in graphics card. Spectators couldn't believe what they were witnessing. It was the most complete computer in existence; no one had ever seen anything like it.

The Apple II was the first home computer to have BASIC built in. It was also the first to have sound, color, graphics, a plastic case (shell), and 48 KB of RAM built into the motherboard without any slots.

The Apple II sold millions of units well into the 1980s. It, along with the burgeoning Commodore PET and Tandy TRS-80, defined the home computer market of the '80s.

PCS IMPACT THE BUSINESS WORLD

In 1979, the Apple II's killer app arrived in the form of VisiCalc. It was essentially an Excel spreadsheet for business users. VisiCalc was originally envisioned as a "magic blackboard." Instead of accountants making one change in a spreadsheet and being forced to recalculate everything (sometimes taking days), now the computer could do it instantly.

When accountants saw the program in action, they exclaimed that their whole job had just become a thousand times easier. There were reports of some individuals crying and shaking at the magnitude of what this software meant. VisiCalc ended production in 1985, but Microsoft picked up the torch with Excel in the same year.

After this, Apple was off exponentially. Jobs was worth over a million dollars when he was twenty-three, and over $10 million when he was twenty-four. By his twenty-fifth birthday, Jobs was worth over $100 million. With the Apple II, Jobs went from working in a garage to being worth over $100 million in just two years.

The future of Apple would not be all smooth sailing, and a PC war involving IBM and Microsoft would take place in the 1980s. We'll be catching up with Apple in the next chapter.

BEFORE DVDS AND CDS, THERE WAS THE LASER DISC (1978)

We're all aware of the compact disc. It's harder to realize what a remarkable invention it was in the post-physical-media age. But when the CD initially came out, it was nothing short of revolutionary. Unlike records or cassette tapes, your music now wouldn't get worn out when you played it back, and the quality didn't deteriorate over time. Sound could technically be stored forever. Digital sound was brought to life by lasers (another AT&T/Bell Labs invention) in 1982. It was the stuff of science fiction. Even as a kid in the mid-'90s, I remember being amazed by the concept.

What many younger people today may also not realize is that the CD was actually based on an earlier technology that was extremely advanced for its time: the laser disc (originally called Disco Vision). I find this odd format fascinating. It should have taken the world by storm, but today it only remains a forgotten quirk of the past.

What Is a Laser Disc?

The Laser Disc was a home-video format, and the first commercial optical disc storage medium.

It was basically the DVD of the late 1970s. There were laser disc players from Phillips, Magnavox, and Pioneer. The year of release: 1978, two years after the introduction of the VHS VCR, and four years before the introduction of the CD. The discs were huge at 12 inches (30.5 cm) in diameter.

The first mass-produced, industrial laser disc player was the MCA DiscoVision PR-7820. This unit was used in many General Motors dealerships for training videos and presentations in the late 1970s and early 1980s. The first movie released on laser disc was Steven Spielberg's Jaws, followed by more than 10,000 titles between 1978 and 2001.

VHS was the main competitor to the laser disc format. The laser disc had the advantages of displaying still images without wearing out a tape on a rotating video drum. It featured a far sharper picture with a horizontal resolution of about 430 lines, while VHS featured only 240 lines. LDs could handle both analog and digital audio, whereas the VHS was analog-only. LDs could store multiple audio tracks—impossible on VHS. This allowed for extras like director's commentary tracks and other bonus features (a first back then, though it is now common for Blu-Ray/DVDs).

Laser discs were initially cheaper than VHS videocassettes to manufacture, as they lacked the moving parts and plastic outer shell necessary for the VHS. In addition, the duplication process was much simpler. (However, due to the larger volume of demand, videocassettes quickly became much cheaper to produce, costing as little as $1.00 per unit by the beginning of the 1990s.) Laser discs also potentially had a much longer lifespan than videocassettes. The discs were read optically, with no physical contact, instead of magnetically. As a result, playback would not wear the information-bearing part of the discs, and properly manufactured LDs would theoretically last beyond one's lifetime.

So, What Went Wrong?

So, if laser disks were so great, why didn't they take off?

The discs were heavier (weighing about half a pound each), more cumbersome, and more prone to damage if mishandled than a VHS tape. Also, the disks could only fit thirty to sixty minutes footage per side, depending on the format, so some movies would require two or more discs. Further, most LD players required users to get up from their couch and manually turn the disc over in order to play the other side.

Error detection also didn't exist at the time. Because of this, even slight dust or scratches on the disc surface could result in read errors, causing various video-quality problems: glitches, streaks, bursts of static, momentary picture interruptions. DVDs and later digital media have such protection.

Despite the mild popularity, manufacturers refused to market recordable laser disc devices on the consumer market, even though the competing VCR devices could record onto cassette. This move hurt sales worldwide. The inconvenient disc size, the high cost of both the players and the media, and the inability to record on the discs, all took a serious toll on sales, contributing to the format's poor adoption figures (though it fared better in the Japanese market).

The last laser disc title release in the US was in 2000, but interestingly enough, players were still being made until 2009.

So, there you have it: the tech of the future that just didn't quite make it to the mainstream. A little quirky piece of tech history.

SONY WALKMAN: A STEREO IN YOUR POCKET (1979)

Sony, who had burst onto the American electronics market with their transistor radio back in the 1950s, were about to have their next big hit: the Walkman. A stereo that could fit in your pocket was the stuff of science fiction, and when it arrived on the market, it completely transformed the way people listened to music.

Sony Walkman

The Walkman was born out of a personal need from Sony's chairman Masaru Ibuka (one half of the hero pair from chapters 7 and 8). He was a big fan of music, but also tended to travel a lot. In order to get his music fix on the go, he would use the existing Sony TC-D5 cassette player. It was portable but not exactly…portable.

The player weighed 1.7 kg and was 237 x 48 x 168 mms (9.3 x 1.9 x 6.6 inches)—doable, but not exactly convenient. Ibuka was a little frustrated with this and thought there ought to be a better way. He assigned an employee, Norio Ohga, to come up with something better. Ohga used existing the tape recorder technology found in Sony's Pressman device to create a tape playback-only model that could be used for music listening. A prototype of the tape media player was cobbled together in time for Ibuka's next flight. It was a crude device, and the custom batteries ran out halfway through the flight, but Ibuka was extremely impressed with the concept nonetheless. He began to champion for a consumer version to be built at Sony.

The other managers at Sony thought it was a terrible idea. Who would want a portable cassette device that could only play back tapes and not even record? What would be the point of such a useless product?

Ibuka strongly argued that listening to your own music while going for a walk would be something everyone wanted. As it turned out, he was right! The result of this little experiment—the Sony Walkman—was a hit. Initially selling for $150 ($500 today), 485 million units were sold in its production run.

The Walkman name remains in Sony's line-up to this day as "high-resolution" music players.

On to the '80s

As the 1970s drew to a close, the world was a different place once more. The computer was now personal, with its own software. It had a growing user base of hobbyists, and even some businesses were using its technology. Two future giants had been born out of this: Microsoft and Apple. We'll take a look at the fierce rivalry between these two companies in the next chapter.

The economy would falter in 1979 with another oil crisis in the Middle East, but from 1980 onwards, every aspect of society would begin to change, irreversibly.

CHAPTER 11

Technology Merges into Society
1980–1989

I've always had an obsession with this decade: its loud clothes and catchy music. Popular films at the time were all creative and original: *E.T., Back to the Future, The Terminator, Ghostbusters, Ferris Bueller's Day Off, Top Gun,* and *Scarface* have all remained modern classics.

You can't think of the '80s without the technology it produced. The charm of the cassette tape hiss and the satisfying click of large physical buttons—it was a time where analog was king.

By the end of the decade, the Western world was on an upswing, but the '80s certainly didn't start that way. It was an economic disaster, involving the worst recession since the Great Depression. In the United States, businesses were failing, crime was up, crack cocaine was destroying communities, and AIDS was a national epidemic. The US inflation rate was almost 14 percent in 1980. Could you imagine everything getting 14 percent more expensive in just one year?

The divorce rates in the 1970s, being higher than ever before, had a devastating effect on the next generation, who had to learn how to live in broken homes. These kids had come of age to the backdrop of Nixon, Watergate, and the Vietnam War. This cynical generation would carry the torch from the baby boomers. They would be known as Generation X.

President Ronald Regan would become a hero to some as he reined in inflation to 3.6 percent by 1983. Conservatism became cool among some young professionals (known as yuppies), who looked down on the hippie idealism of their parents.

The Soviet Union was still a force at the start of the decade, and Japanese innovation was on the rise, taking over American roads and electronics stores. Japanese products, once seen as low-quality and cheap, were not anymore—they proved themselves during the decade. Detroit couldn't keep up with the well-made, cheap, and fuel-efficient cars. In 1980, the car manufacturers in Detroit lost $4 billion.

During the worst years of the '80s, a creative boom breathed life into despair. In New York, hip-hop and breakdancing would find their way into the mainstream, while in California, a new culture was beginning to form around skateboarding, with a young Tony Hawk taking the lead. Hawk pioneered acrobatic aerial tricks and, thanks to the phenomena of easy-to-use handheld VHS recorders, shot a film in 1984. It would create the foundation for a new sport.

NASA, which had been created in response to the Russian threat of *Sputnik* and other space missions, reached its zenith in the 1980s with the space shuttle program. However, the pinnacle of American pride was postponed for a number of years after the Challenger disaster in 1986 was broadcast live on CNN. The disaster was caused by a frozen O-ring that led to the separation of a solid rocket booster, resulting in a catastrophic structural failure. All seven astronauts on board lost their lives.

Electronic spreadsheets, such as VisiCalc and subsequent software, revolutionized the business world and the rest of society in the 1980s. Business scenarios could now be played out in real time, and profits and losses could be known in seconds instead of days or weeks. The stock market saw a major boom due to computerization. The decade also saw a turn toward individualism: it seemed that everyone was making money hand over fist. The movie *Wall Street* and Gordon Gekko's statement "greed is good" summed up the attitude of the latter part of decade.

In the 1980s, the technological innovations of the '60s and '70s started to become useful in everyday ways. VHS and cassettes became popular. CDs had their time in the spotlight at the end of the decade. The Walkman made every user the star of their own movie, with a personal soundtrack setting a virtual mood as they walked around.

The kings of technological innovation in the 1980s were the United States and Japan, but a less-recognized PC market would also flourish in the UK. By the end of decade, a world of connected information— just as the great visionaries of the 1950s and 1960s imagined— was established.

In this chapter, we'll see how technology becomes a force that takes over the modern world; but first, a little ear and eye candy.

I WANT MY MTV! (1981)

MTV would launch on August 1, 1981. At the time, it was a new form of entertainment: a media video jockey, available right at home. The channel would define a generation—the "MTV generation."

The first video was "Video Killed the Radio Star" by the Buggles—somewhat of a prophetic take on how the channel would change the music industry.

Though MTV seemed rebellious, its roots were purely corporate. The television company, Warner Cable Corporation, already had two main channels in 1980: The Movie Channel and Nickelodeon. A channel focusing on music seemed to be a good next fit. Warner originally rejected a proposal for a twenty-four-hour music channel that was, as the directors saw it, essentially radio with pictures. After some internal struggle, it was decided that music videos would be cheap to implement because the record companies would provide the content for free and, in return, they would receive exposure. MTV had to be created in just seven months, as Warner wanted it ready for the summer, when college students would be on break. The initial run of MTV only had 250 music videos and the station was, surprisingly, only available in smaller American towns.

Arguably, the first band to make it big because of the new medium was Duran Duran, who boldly released an entire album's worth of movie videos as a kind of short film. The stunt got people talking. Soon every band wanted to make videos that would hit it big on MTV.

Despite this, MTV was initially a financial disaster. The station approached Mick Jagger of the Rolling Stones, which was a high-profile act at the time and demanded a hefty fee. In a meeting with the rock star, one of the producers (perhaps as a joke) threw down a $1 bill on the table and said, "That's all we can afford." Jagger thought this was a ballsy move and agreed to do the gig for $1. In 1982, the famous "I want my MTV!" campaign starring Jagger was born, and soon, TV stations around the US were being flooded with phone calls asking for the new channel. Within a few years, MTV was in every state, forever changing the landscape of the music industry. As a kid, I remember the rare times I had the chance to see MTV in Australia. I thought it was the coolest concept ever. It captured the feeling of what was new and what trends were around the corner.

Pac-Man! (1980)

Pac-Man is one of the most popular games of all time. The game was released in Japan by Namco in 1980 and made its way to the United States in the same year. It was one of the first major games that wasn't based on space or sports.

The game was the brainchild of twenty-five-year-old Toru Iwatani, who took the game from idea to reality in just one year.

The game was originally named Puck Man, but the Americans thought that it sounded too much like...you know. The worry was that gamers would strategically scratch out part of the P, so the name was changed to *Pac-Man*. The "power-up" idea was inspired by Popeye, who would eat spinach and gain superhuman strength.

The game was a little weird, so it wasn't expected to be a hit...and, at first, it wasn't. But soon popularity grew, and by the end of 1980, *Pac-Man* had made more money than *Star Wars* (the highest-grossing film at the time). The game was played more than ten billion times in the twentieth century. Today, *Pac-Man*'s iconic shape is instantly recognizable around the world.

The Video Game Crash of 1983

Released in 1977, the Atari 2600 home console—hot on the heels of the Magnavox Odyssey (chapter 10)—was the king of the home-videogame market. By 1980, Atari was the fastest-growing company in United States history, going from $70 million to $2 billion in sales in a few short years.

It wasn't long before other companies took notice of Atari's success. Video games were a hot commodity, and everyone wanted to get in on the action. Some firms even switched industries to do so. It would spell disaster, and the new video game industry almost ended before it started.

The video game market quickly became over-saturated with consoles and games, and yet still more were being made. All of these consoles were also competing with the newly emerging personal computer technology. PCs were better value, as they could do more than just play games, further harming the demand for standalone gaming systems.

To make matters worse, the people who actually *made* the games weren't getting credit. Even if a game went on to sell millions, the programmer would only make a flat wage of roughly $30,000 per year. A few frustrated Atari game developers became tired of this arrangement and founded Activision—an independent company that would make third-party games that were compatible with Atari hardware.

This was a disruptive change to the video game production model. Soon, there was a tidal wave of independent game developers. To give you an idea of how bad things got, food companies like Quaker Oats and General Mills started making their own video games. What did food have to do with video games? Apart from Doritos dust on the controller, absolutely nothing. At this point, anyone could make an unauthorized game for an Atari or any other system. It was a fatal flaw that would eventually help bring down the industry.

Atari Brings It All Crashing Down

Ironically, one of the first cracks in Atari's foundation came as a result of the industry's most popular games. *Pac-Man* was a huge hit in the arcades, and Atari wanted in. The company bought the rights from Namco in 1981. But, as the year was almost over, and management wanted the *Pac-Man* port (Atari version) to be done by Christmas, Atari rushed out the game. The result was a disaster. The Atari *Pac-Man* was slow, full of bugs, and didn't even resemble the original game. The result was five million unsold cartridges.

Despite this setback, Atari was still king of the industry, and the health of the industry as a whole rested on its back. But in 1982, Atari would make the same mistakes. Again, they had a big franchise, Steven Spielberg's *E.T.*, and again they rushed production. This time, they only gave their programmer six *weeks* to create the game. The result would disgracefully become known as the worst video game in history, and another five million copies were left unsold in the warehouse.

The 1983 earnings statement for Atari showed 10 percent growth instead of the forecast 50 percent. Wall Street didn't like what they saw, and stock investments in video game companies crashed. This event sent a shockwave through the industry, bringing it to its knees.

By the end of 1983, Atari was $500 million in debt.

As for the ten million unsold Atari game cartridges? Twenty truckloads' worth were driven to a landfill in New Mexico and buried. Interestingly, they would be partly excavated in 2014.

Video games were largely written off as a one-time fad, but a few savvy Japanese businessmen thought that maybe video games *did* have a future. Their company would be called Nintendo.

Nintendo Entertainment System: NES (1985)

The Nintendo company dates back to the late 1800s, when they distributed card games. In the twentieth century, they switched to toys. In the 1970s, seeing Atari's success, the company would make *Pong*-clone home consoles. In 1980, they would release the handheld Game & Watch which, like the name implies, was a small (calculator-sized) game that also contained a clock. Next, the company attempted to make cabinet arcade games, which were largely failures. The CEO of Nintendo USA, Minoru Arakawa, asked his entire company for the next big idea, hoping to stumble across something interesting. A twenty-nine-year-old artist, Shigeru Miyamoto, was up to the task.

Nintendo had hoped that basing a game on the character Popeye (who was popular at the time) would boost sales. However, Nintendo couldn't get the rights, so Miyamoto altered the characters slightly. Olive Oyl became a princess (Pauline), Popeye became "Jump Man," and Popeye's large and dopey enemy, Bluto, became a gorilla. The finished game was called *Donkey Kong*.

Released in 1981, *Donkey Kong* became a hit. 60,000 arcade machines were in the US by the end of the year, earning Nintendo $180 million.

Buoyed by this success, Nintendo would release the Famicom (short for family computer) in Japan in 1983. Early titles for the Famicom were ports of Nintendo arcade titles. The company had its eyes set on North America, but after the video game crash, US stores were wary.

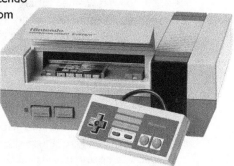

Nintendo Entertainment System (NES)

In 1985, the console was repackaged and rebranded for the US. It came with a light gun and a robot named R.O.B, and the whole package was renamed the Nintendo Entertainment System (NES). The plan was to distance the NES from the concept of video games. It didn't work.

Finally, Nintendo-Japan CEO Hiroshi Yamauchi decided that, if Nintendo was going to sell in the US, his team would have to go there themselves. A "Nintendo SWAT team" was deployed to display the game and get the product onto shelves. The team even offered stores a no-risk proposition. Stores only had to pay for the products they sold; if the NES didn't sell, Nintendo would take the systems back. The NES ended up being a smash hit.

In 1985, "Jump Man" (now called Mario) featured as a standalone on the NES. The game was like nothing anyone had seen before: a vast and imaginative world with smooth graphics and side scrolling visuals. This of course was *Super Mario Brothers*. Forty million copies of the original game have been sold to date. I remember playing *Super Mario* in the '90s, and it is as much fun now as it was then. This success would be followed by a second smash hit in 1986: *The Legend of Zelda*.

Learning from Atari's mistakes, Nintendo closed off their system to outside developers with a special in-house hardware chip. This meant they had complete quality control over what games could be played on their system. This time, third-party developers introduced great titles that only helped propel the console's success.

In ten short years, Nintendo had gone from not even being in the video game industry to owning the home console market.

Facts about Mario

- Mario was named after Mario Segale, the landlord of Nintendo America's office, who barged in on a company meeting demanding overdue rent.

- Mario has appeared in over 200 video games so far, which together have sold over 193 million units. *Super Mario Brothers 3* alone had made over $500 million in the US by 1993. So, it's safe to say he's a big deal.

- Mario doesn't have the hat and moustache because he is Italian. Shigeru Miyamoto drew Mario wearing a cap because he found hair difficult. He also drew the moustache because it was easier to see than a mouth in pixelated video game graphics.

BOOM BOXES (EARLY 1980S)

The transistor radio combined with a tape deck and huge speakers turned out to be just the thing the youth of the 1980s wanted. The boom box (or ghetto blaster) was big business and Sony, Panasonic, and General Electric were some of the biggest names in the game. A key feature was to have an input for microphones and turntables, making it a makeshift all-in-one live performance system. Another key development was speaker bass, the more of it, the better. Kids wanted to feel the beat in their chests as they slammed down their lyrical creations or threw it down on the floor.

The boom box became popular with inner-city youth, and it played a key role in the rise of hip-hop music and culture. The boom box also spawned a new kind of dancing on the streets. Kids would chain together energetic moves, literally spinning on their heads at times. The style was originally called B-boying but would later become known as breakdancing.

A Boombox Tape Deck Kick-Starts Rap

The portability of the boom box allowed for the hottest underground sounds to be recorded on tape and shared with others. It was like a 1980s Spotify. Bootleg copies of New York DJs like Grandmaster Flash and Kool Herc would be traded on the streets. These tapes would later inspire a whole other generation of rappers like Run DMC, LL Cool J, Public Enemy, the Beastie Boys, and Eric B & Rakim, who would take hip-hop into the mainstream.

New Thinking: The PC Assists Humans

In the 1980s, computers began to mature and transition from the toys of nerdy hobbyists to a professional consumer product. The biggest names of the decade were Commodore, Apple, IBM, and Atari, each of which we'll look at below.

Commodore Computers: A Silent Success

Commodore is a forgotten name in computing today, but they were once a big-name PC manufacturer. Released in 1982, the Commodore 64 still holds the record for the bestselling computer of all time, selling 17 million units from release until its discontinuation in 1994.

Another notable design was the Amiga line by Commodore. I find this model fascinating because it was so far ahead of its time. In a world of (mostly) monochrome, Amiga computers featured color graphics and all types of multimedia due to custom chips. Bad management kept the computer from reaching its full potential. However, at the time, better graphics and sound could have held the Amiga back in sales, as business computers had to be "serious" and clearly defined for work only.

The PC is the Amplifier of Human Ability: Apple

In a 1980 speech to investors, a shaggy-haired Steve Jobs would give his vision on the future of computing, long before most others realized its significance.

Jobs recalled reading a 1973 issue of the *Scientific American* as a kid. This particular issue featured a study on locomotion. The magazine ranked the efficiencies of different animals in terms of how much energy they required for movement. The condor won by a vast margin. A walking human didn't do so great, placing about a third of the way down the list; but here's the twist: a man with a bicycle was twice as efficient as the condor.

This lit a light bulb in Jobs' head.

The study showed Jobs that man was an exceptional tool-maker, yet there was a deeper message than that. It's fair to say that the Industrial Revolution, examined at the start of this book, was basically an amplification of human physical ability. Sweat was replaced by steam, allowing man to do exponentially more and use power much more efficiently. In the 1980s, personal computers allowed for a similar efficiency and furtherance of human mental ability. Jobs would call the computer the "bicycle of the mind."

December 12, 1980 was the day Apple first went public. It was the biggest IPO since the Ford Motor Company in 1956. Despite this success, the hippie inside Jobs remained. The twenty-five-year-old would often walk around the office nibbling on a bag of seeds. He sometimes didn't shower, and often went around barefoot. By this time, Steve Wozniak was taking a less direct role in the management of the company.

These were good times for Jobs and Apple, but the barefooted hippie wouldn't stay unchallenged for long.

THE BLUE SUITS ENTER THE PC MARKET: IBM

When Apple arrived on the scene, IBM was the largest technology (but not PC) company in the world. IBM's spirit couldn't have been any more different from the Silicon Valley countercultural spirit. IBM's employees were traditional, straight-edge college graduates. There was even a strict but unspoken dress code of blue suits and white shirts. IBM was all about large mainframe computers and the old way of doing things, while the PC market seemed too new, wild even—fueled by passion and curiosity.

In 1980, IBM was watching Apple's success in the PC market and began to realize they were wrong about the PC. It could actually be used to extend human intellect, and even for communication. They also saw dollar signs, as the PC industry was already worth $1 billion just three years after it had begun.

The problem was that IBM was bureaucratic and very slow to move when it came to decisions. It would take years just to decide on a PC design—how would they ever come up with a PC product in time to compete?

To speed up the process, a small team within IBM decided to use off-the-shelf parts for their first PCs. By not building the computer from scratch, they managed to get a business-oriented personal computer built in just one year. Using non-IBM parts was a very unusual practice for IBM, but with the hardware cobbled together, the only thing needed was software to run on it.

An operating system (OS) is a computer's digital "traffic cop," keeping track of how files are stored and telling the computer how to handle hardware such as a mouse, screen, or floppy disk drive. BASIC already existed, but it was only a rudimentary user interface. IBM needed more if their computer was going to take over the world.

This is where things start heating up. There was a battle brewing across the computer industry for the operating system that would rule them all. If the '70s was a time of love and sharing in computing, the '80s was a time of theft and exploitation.

Herein lies one of the saddest stories in computing history.

Bill Gates vs the Man who Could have been Bill Gates

While building their new PC in 1980, IBM approached Microsoft to build the operating system. Before saying a word about the project, the boys in blue asked Bill Gates (who was initially mistaken for the office intern) to sign a non-disclosure agreement. Gates had to keep IBM's plan a secret. When Gates was told about the plan for Microsoft to supply an OS (operating system), he remarked that Microsoft didn't have one. Instead of creating something he didn't have, like he did for MITS and the Altair, the twenty-five-year-old programmer directed IBM to renowned computer scientist Gary Kildall.

Kildall, who sported a red beard, was a mild-mannered man, but despite being unassuming he had already significantly impacted personal computers. In 1971, Kildall had explored making a programming language for Intel's first CPU, the 4004, but he realized that it needed a system to control how the chip interacted with the rest of the processor. In 1972, this system became CP/M (Control Program for Microcomputers), the very first general-PC operating system. This was different from Microsoft's BASIC, which was an only surface-level program that ran on top of a computer's operating system (the part that can tell the hardware how to behave).

Before CP/M, each computer had to have a tailor-made operating system. It was like having different types of fuel for every car model in existence. With CP/M, one only needed to write software once and CP/M would take over the rest. It was a way to run the same software on different computers with little effort.

Kildall didn't have much interest in business matters, but his wife, Dorothy, convinced him to start licensing his creation. The company Digital Research would be founded to do so. By 1979, Digital Research became the industry standard for operating systems. In essence, they were (today's) Microsoft of the late 1970s, and Kildall was the era's Bill Gates.

The Worst Day of Gary Kildall's Life

Back to the story. Keen to waste no time, IBM took Gates' advice and paid a visit to Kildall in Seattle. Gates called Kildall to warn him that someone was coming to visit, exclaiming, "Treat them right; they're important guys!" Because of the non-disclosure agreement, Gates couldn't reveal exactly who was coming to visit. Unfortunately, Kildall didn't fully understand Gates' warning. Perhaps he thought it was just another small company, not the largest tech company on the planet.

Whatever the case, Kildall wasn't home when IBM visited. Although the exact details are murky, according to Jack Sams (ex-IBM CEO), Kildall was out flying one of his private planes on business.

The blue suits ended up talking to Kildall's wife, who was now head of operations at Digital Research. IBM's lawyers began pushing her to sign a non-disclosure agreement to essentially say there were never there. Dorothy Kildall wasn't impressed with this and refused to sign the document. The IBM team had a short temper—time was money! After getting nowhere in negotiations, the suits became frustrated and left Kildall's house.

A few days later, IBM approached Bill Gates a second time. This time, Gates took the opportunity. He saw that IBM had the potential to change the PC market, with a cleaned-up business image. Here comes the clincher: You know how I said that Gates wasn't going to sell something he didn't have yet, this time around? Well, that idea quickly went out the window. Gates told IBM that Microsoft could in fact make an operating system, even though they didn't have one.

Microsoft would cheat and buy an operating system from a small company, Seattle Computing Products, down the road. Microsoft would pay $75,000 for the OS made by Tim Paterson of Seattle Computing Products. This operating system was called the "Quick and Dirty Operating System," or QDOS. If you think that's a weird name, there's a reason for it. The code for this software was essentially a rip-off of CP/M, Gary Kildall's software! It did have a few differences but was suspiciously similar.

Seattle Computing Products had ripped off CP/M and Microsoft now had its hands on a functioning operating system. QDOS would become Microsoft DOS, or MS-DOS, and would be packaged with every IBM PC. So, if you've ever used MS-DOS, you could say you were using the "Microsoft Dirty Operating System." (Its name would deceptively be changed to "Microsoft Disc Operating System," officially.)

The first IBM PC was released in August of 1981 and was predicted to have 250,000 sales. It ended up selling 2 million units in a couple of years, overtaking Apple as the world's largest PC manufacturer.

Bill Gates was about to become the richest man on earth. The dirty operating system was half of the equation; it would be smart business sense that completed it.

Because the IBM PC was made from off-the-shelf parts, other manufacturers such as Compaq and HP began making their own PC clones with the same parts. Microsoft had licensed MS-DOS to IBM for a one-time fee of $50,000—but there was a catch. Microsoft never mentioned to IBM that their deal was non-exclusive. Soon Microsoft was selling MS-DOS to all of IBM's competitors, taking with it a licensing fee from every computer sold.

This licensing deal has been called the greatest (or perhaps most cunning) deal in the history of the world. It made Bill Gates a billionaire. Kildall's failure is conversely called one of the biggest business failures in all of human history. After realizing what he had lost, Kildall would shed his kind nature and threaten to sue IBM.

In a settlement, IBM agreed to offer a version of Kildall's CP/M software, alongside MS-DOS, with every PC they sold. Strangely enough, Kildall was pleased with this: in his mind people could now choose for themselves which software they liked best. Justice had finally been served. There was only one problem...when IBM brought out the two software choices, MS-DOS sold for $40 and CP/M for $240!

IBM had tricked Kildall again, and Kildall's CP/M software would fade into obscurity, losing to a clone of itself. Kildall didn't take it so well; he was so crushed by the events that he didn't even bother suing Microsoft or IBM again. The strain from missing out on the greatest opportunity of many lifetimes probably aided Dorothy Kildall to later file for divorce. The ubiquity of personal computers in the following years meant that Kildall would be reminded of his failure everywhere he turned. Kildall slipped into a bout of depression and alcoholism. In 1994, Gary would die from head injuries after a fistfight at a biker bar.

Today, Kildall is only a footnote in technology history, so it's important we keep his contribution to the evolution of computing alive.

Steve Jobs "Thinks Different"...by Borrowing

Meanwhile, Steve Jobs set his eyes on a new problem. Before anyone could solve their problems with a personal computer, they needed to learn how to use one. This was a big hurdle in the '80s. Typing text commands wasn't intuitive to the average user. What was needed was an easy way to interact with a computer: a way that seemed natural.

In a 1980 speech, a bearded, skinny, long-haired Jobs remarked that it was incredible that the power of computers had already increased exponentially in just three years. People would ask him all the time why computers needed more power. Jobs' reply was pretty insightful:

"We're going to use this power to go back into solving the problem of making computer interaction go more smoothly. That is, making computers easier to use and specifically not for number crunching." Jobs knew this with such confidence because, just a year earlier, the PARC establishment had shown him the Xerox Alto. Remember that computer that was basically a desktop PC, teleported from the future? Yep, that one.

Xerox Alto: The Day of Failure

Xerox Alto technology would be used to create the Apple Lisa, whose technology trickled down to the 1985 Macintosh, whose ideas influenced Microsoft Windows. The latter two are conceptual ancestors of the touch-based devices we use today.

So, let's get back to the story of how Xerox let the computer of the future get stolen by Steve Jobs. As you'll recall, Xerox (of PARC) invited a twenty-four-year-old Jobs over to their research institute in 1979, to see if he could help them reduce costs of production. The deal saw Xerox gain a million shares of Apple stock, in exchange for Jobs getting inside information about everything cool and revolutionary going on at PARC. Nobody checked with the guys at Xerox development—business development signed off on it anyway.

At the showcasing of the Alto, seeing the future of the computer before him, Jobs demanded his entire programming team get to see the demo. There was protest from the Xerox PARC staff, but upper management wouldn't budge, and granted Jobs his request. Xerox had unknowingly just given up the kitchen sink and the future. For

Jobs, it was all fair game. His reasoning? A quote from Picasso: "Good artists copy, great artists steal."

The first Apple product to integrate this new way of thinking was the Apple Lisa in 1983.

As we saw, the power of the Alto (and subsequently the Lisa) was its graphical user interface. Before this innovation, computers were complicated and daunting. If you wanted to use a program, you had to know the right commands and type everything by text.

For example, if you wanted to copy a file called "Darude-sandstorm. mp3" to your music folder, you'd have to type EVERY single letter of this phrase: "copy c: \file Darude-sandstorm.mp3 d:\music" (mp3s hadn't been invented yet, by the way).

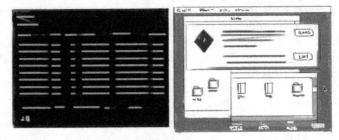

Text based interaction vs graphics on the Apple Lisa
(borrowed from Xerox)

Instead of memorizing a long list of commands to navigate within a program, the Lisa—based on the Alto technology—would just show a graphical drop-down menu. All you had to do was click on the option you wanted.

The personal computer was now simple and intuitive: just drag the mouse and click to interact with the files directly. Pictures instead of just words. This was the power of the graphical user interface.

Umm, Steve...? Apple Needs a *Real* CEO

During the Lisa's development, and following Apple's IPO, Apple executives asked Jobs to bring in someone who could balance his idealistic mentality with a hard-numbers temperament. Jobs was thought to be too volatile and inexperienced to head the Lisa project and was banned from working on it. In 1983, Jobs would have a meeting with PepsiCo CEO John Sculley. Sculley was a straight-edge cookie-cutter businessman. He seemed perfect for the role and was hired.

Jobs was a little disheartened by being banned from the Lisa project, but quickly turned his attention to a small research team within Apple. The team had just built a small all-in-one graphical computer. Seeing this, Jobs asked to be head of the team and was appointed. The low-cost graphical would be called the Macintosh (named after an American apple) or Mac for short. It had to be great, because the clock was ticking—by 1983, the IBM PC was picking up steam. The Blue Suits were on the move!

THE ORIGINAL APPLE MAC

The Mac was a new take on computing aimed at opening up the imagination. It was friendly and personal: a window into art, education, and business. It was more than just a machine.

The designers of the IBM PC, on the other hand, had taken a corporate angle. Ugly, cold, and clunky, the machine was simply a tool to get work done.

Production within Apple was divided into teams: the Mac team was young, with an average age of twenty-one. They were creative types: musicians, poets, and artists, yet also great computer scientists. The Mac team would fly a pirate flag above their office space to show they were renegades. This team would be very competitive with the Lisa and Apple II project teams—a somewhat unhealthy rivalry behind company doors.

Although strong on paper, the final Lisa product wasn't a hit. The Lisa got bogged down in numerous design problems and ended up taking 200 collective man-years of development. Additionally, it was too expensive at $10,000 ($23,000

today), and people couldn't understand what the hell it even was. It flopped.

In contrast, when released in January 1984, the Mac initially got off to a flying start; however, sales began to run out of steam in a big way.

By 1985, all the Apple enthusiasts had bought a Mac, yet disaster was looming. The Mac critics began to see the machine, not as a productivity tool, but as a toy. They were kind of right. There wasn't a lot of software for the Mac even a year after its release, because Jobs' team failed to finish the release of Macintosh Office. Mac Paint and Mac Write were the only two major pieces of software on the Mac, and it still cost $1000 ($2300 today) *more* than the IBM PC. Further, the Mac was short on RAM due to its graphical interface, so whenever you wanted to change programs, you had to switch floppy discs.

At this stage, the PC was outselling Macs twenty-to-one due to these issues. Many analysts were saying Apple would be over within a couple of months. Even the trusty Apple II was falling in sales in comparison to the newer IBM PC. Jobs was emotionally invested in the Mac and didn't believe the numbers. He acted like the Mac was the hottest-selling computer of all time, even though it was doing just average. To Steve, it didn't make sense: *How could the computer of the future not be selling?!*

During all of this, Bill Gates of Microsoft had been given access to the Mac for two years to develop software for it. Apple had discounted Microsoft as competition, because they had their eyes focused on IBM. Steve Jobs didn't know the opportunistic nature of Gates, and this brief partnership proved to be a fatal mistake for Apple.

While working with the Mac, Bill Gates realized that the graphical nature of the Mac was going to be a threat to his clunky Microsoft DOS—a text-based operating system. He needed his *own* version of the graphical future. So, Gates' team simply built a graphical interface on top of DOS. They called it Windows 1.0.

It was buggy and not pleasant on the eyes, but Windows was still graphical and that was enough to keep MS-DOS in the game. This first crack in Apple's graphical fortress would eventually grow.

When the Mac weren't selling as well as Jobs had envisioned, the relationship between Sculley and Jobs began to sour. Sculley thought money should be poured into the Apple II to extend its life, but Jobs disagreed. He wanted to push for a windowed, graphical future with the Mac.

In 1985, the executive board at Apple was called in to decide the fate of Jobs. Due to the conflict between Sculley and Jobs, one had to go. The board was split. On the one hand, a lot of them appreciated Jobs; on the other hand, Sculley had experience in leading corporations and could be just the change Apple needed. In the end, the board chose Sculley. During a later interview, Jobs would solemnly remark that John Sculley "destroyed everything that I had spent ten years working for."

Following Jobs' exit, the Mac was saved by the Adobe company. Adobe created great software for the Mac and birthed desktop publishing. By 1987, Macs were selling one million units per year, very close to what IBM was managing—all without Jobs.

I have some great memories of the color Mac, as my backward primary school still had a set of these in the computing area in 1998. Two of my favorite games were *Crystal Quest* and *Q*Bert*. Just thinking about it makes me want to download copies right now.

Steve's Wilderness Experience

After being more or less fired from Apple, Jobs took some time away to reflect on himself and, in doing so, realized his immaturity. He would stop chewing seeds, put on some shoes, and morph into a true professional. It's often said that there were two Steve Jobs: the one before getting fired, and the one after.

In 1985, Jobs would entice a loyal team of Apple employees away and begin on his next venture…NeXT.

NeXT was to be a cutting-edge computer hardware and software company. The team got to work on building an operating system called NeXTSTEP for their new computers. These computers were expensive, starting at around $6,500 ($14,000 today).

In 1988, NeXT's aim was to compete with mainframes and Unix workstations, by being small yet powerful. NeXT's computers were so powerful for the time they claimed they were "a university in a desktop." They were supposed to be able to store large quantities of information with advanced search. They were also built to be powerful enough for complex problems at a university level.

When asked what the purpose of such a machine was, Jobs would remark: "I want some kid to figure out how to cure cancer in his dorm room on one of these things."

It sounds like a strange quote, but there was solid reasoning behind it. While on tour promoting the Mac to universities, Jobs met a Nobel Laureate chemistry professor who had a frustrating problem. The professor was trying to teach his students about forming DNA in a laboratory, but things weren't going as well as they could have.

The students were learning from textbooks, as physically doing the experiments was too expensive for undergraduates. The professor knew that computers could simulate the forming of DNA, but that was the stuff of supercomputers, not personal computers. A request was made to Jobs to create a small, powerful computer with a megabyte of RAM and a megapixel screen tailored for universities.

This was the germinating idea that would result in the NeXT computer, which would go on sale in 1988 with 8 MB of RAM, a 256-MB hard drive, and a 25-megahertz processor.

Aside from the aforementioned qualities, NeXTSTEP software made it easy to build apps—especially the graphical interfaces of apps. Does an app need a text box to input a user's name? No worries, just drag and drop it in. Need some buttons? Just drag and drop! It was pretty advanced for the day.

Jobs used to remark that an app that took twenty programmers two years to make on Microsoft's DOS, would take only three programmers six months on the NeXTSTEP system. While the Mac brought graphical user interfaces to computers, NeXTSTEP brought modern design techniques to app building. Its simplicity would play a role in the foundation of the World Wide Web (more on this later).

In 1989, Jobs forecast 10,000 NeXT Computers would be sold each month. In preparation, he built the most automated computer factory on earth, with a capacity of 150,000 units *a month*. The *total* amount sold was a modest 50,000, up until production ended in 1993. NeXTSTEP computer hardware may have faded, but the NeXTSTEP OS remained. The versions of NeXTSTEP that were ported to other hardware became known as Openstep.

In 1997, Apple would buy the Openstep OS from Steve Jobs and, as a consequence, he received a seat on the board. From this point on,

Jobs managed to climb his way back to the top of Apple and officially become CEO again in 2000.

As for NeXTSTEP, it became the foundation of a redesigned Mac OS (Mac OS version ten) and the basis for iOS, the operating system used in the iPhone.

THE ORIGINAL PURPOSE OF THE WORLD WIDE WEB

The visions of Vannevar Bush, Licklider and Engelbart have all led up to this moment. In the 1980s, the internet existed to share data, but it wasn't revolutionary in this basic form. It was only being used in closed networks and getting from one place to another was no easy task. It would be the World Wide Web, which was built on top of the internet, that unleashed the full potential of exponential knowledge growth and human progress. Here is the story of how the Web began.

In 1980, 10,000 people were working at CERN: that's the place in Switzerland with the Large Hadron Collider (a particle accelerator used to discover new particles) and other goodies. There were countless hardware and software requirements at CERN, and email was the main form of communication. However, it was a mess: when different projects began to collaborate, there needed to be a way for scientists to keep track of everything: results, research teams, phone numbers, etc.

By 1984, most of the project-coordination work was being tied up in updating information, and as people came and left, information was being completely lost. Additionally, text files rarely worked from one computer to the next.

CERN called in British computer scientist Tim Berners-Lee to help fix the problem. Berners-Lee had been messing around with hypertext (the innovation of Engelbart we discussed earlier, in this context used to link from one document page to another), and developed a program called ENQUIRE. However, ENQUIRE didn't work as expected. The issue was two-fold: The full collection of pages in the company weren't available to everyone and ENQUIRE only allowed people to jump between pages that had the same categories, e.g. just staff members, or just a set of results.

To fix this, Berners-Lee changed how information at CERN was accessed: the full set of ENQUIRE pages should be accessible to *everyone* but stored on the network to avoid compatibility issues. Each link should link to *any* other page, regardless of what category it was in.

Here was Tim's new thinking: In the physical world, people may be nicely arranged in a management structure, but the information inside that management structure and beyond it aren't arranged so neatly. If you let people who are connected share any information they like (regardless of what it's about), the relationships between different types of information are messy and dynamic. However, this *does* in fact reflect our natural way of associating: if you were to draw the connections and relationships between these pieces of information their shape would be like a web—chaotic yet connected. If you remember back to Vannaver Bush's ideas of the Memex in "As We May Think," you can see the similarities in linking information by mental association.

Arranged in a web, HyperText was perfect for modelling the changing and chaotic relationships at CERN. The whole management system could grow and adapt in a natural way. The World Wide Web owes its existence to HyperText. If you don't believe me, HTML stands for **H**yper**T**ext **M**arkup **L**anguage, and HTTP stands for **H**yper**T**ext **T**ransfer **P**rotocol. The internet would be pretty much nothing without it.

Berners-Lee wrote a proposal on March 13, 1989, for "a large HyperText database with typed links." Although the dry proposal title attracted little interest, Berners-Lee was encouraged by his boss, Mike Sendall, to begin implementing his system.

The Early Web

Within a year, Tim's HyperText database project was up and running, but no one wanted to use it. To get people to use it, Berners-Lee put the all staff phone numbers on the database. It was sneaky, but it worked. This CERN phone book is now a piece of history and is one of the earliest webpages on the internet.

Not long afterwards, Berners-Lee gave a presentation about the web, on the web, which, amazingly, is still on the web! (You can find it here: https://www.w3.org/History/19921103-HyperText/HyperText/ WWW/Talks/Title.html.)

Soon others outside of CERN started putting their own information on the internet by using the web. A few years later, it became the greatest explosion of information in all of history, propelling information faster than ever, and creating breakthroughs in all areas of knowledge through instant global communication. The original vision of Vannaver Bush was now complete.

Berners-Lee considered several names for his innovation, including *Information Mesh*, *The Information Mine*, or *Mine of Information*. However, he settled on *World Wide Web*. So, instead of saying "Go to WWW..." we could've been saying "Go to IM or MO..."

TECHNOLOGY ENTERS THE MUSIC INDUSTRY: ATARI

I'm obsessed with music from the 1980s, especially from later in the decade. As a music producer, I'm intrigued by how the sounds of '80s electronic music scene were produced given such limited tools.

Atari ST (1985)

The Atari ST was a powerhouse, offering more than the Mac did for a lower price. It was the first computer to have one megabyte of RAM for less than $1000. It was also instrumental in the rise of electronic music.

It allowed enthusiasts, perhaps for the first time, to easily compose music on a computer, using software like Cubase or Logic.

Upon its release in 1989, Cubase was one of the first pieces of professional production software that allowed for the graphical arrangement of music. Other music software would use text-based functions.

The monochrome screen of the Atari ST was said to be so easy on the eyes, it was almost like looking at paper. The ST was fortunate to have extremely low latency, even when compared to professional equipment. Low latency means less lag: getting a sound as soon as you press a button or note. Artists like Fatboy Slim, Madonna, Aphex Twin, and Darude (of Sandstorm fame) all used the Atari ST at one point.

Around this same time, synthesizers were also becoming popular, transforming the soundscape of pop music. Early examples include "Tainted Love" by Soft Cell, and "Sweet Dreams" by the Eurythmics.

The Roland TR-808

In 1980, the Roland corporation released the TR-808 drum machine, which would become known simply as the 808. When it was first released, it was a commercial failure. Producers and musicians wrote it off as rubbish, citing that it sounded nothing like a drum kit. It was too synthetic, too electronic.

Some people did find the sound appealing. Its strange, synthetic, crazy hard-edged sound was something not heard before. The 808 was the first drum machine to allow custom rhythms instead of presets. The machine was easy to use—it would continuously cycle through a bar of music and you would press a button to add the beat you wanted. Essentially, it required mucking around and coming up with sounds you wouldn't imagine creating yourself. In addition to this, all the frequencies output by the machine seemed to be designed in *just* the right way to capture the human ear's attention.

In the early 1980s, Chicago radio DJs were playing various styles of dance music, including older disco records, but also emerging hip-hop music. A few DJs took things a step further—they would make mixes from records on the fly with reel-to-reel tape. Drum machines added the last piece to the puzzle. The hard edge of the 808's sound became synonymous with Chicago and Detroit.

The music quickly spread. A DJ would pass a tape to a friend, and that tape would get copied and given to a few other people and before you knew it, a song could be heard all over Chicago. The sound became known as "house." It was the biggest musical phenomenon since rock and roll in the late 1950s.

Competition between DJs started happening and in 1985, the first house record came into being. It was "On and On" by Jesse Saunders, and it heavily used the 808.

By 1985, every kid in Chicago with a drum machine was making house music. They were innovators and pioneers. These kids figured out how to use drum machines and synthesizers without being musicians. The ease of use played a pivotal role, and soon the 808 was the Fender Stratocaster of electronic music.

The machine has a lasting influence till this day. The 808 snare, high hat, and kicks are used heavily in trap music today. Kanye West even paid homage to the machine with the album *808's and Heartbreak*.

The Roland drum machine and bass synthesizer influenced house in Chicago, the techno scene in Detroit, and electro in New York (hip-hop), and later affected the entire world.

The harder Chicago sounds traveled to the UK (first the north in 1986, and then the south). Machines like the Atari ST were key to the composition of these early tracks.

By the summer of 1989, thousands of people were attending commercially-organized underground house music parties, called raves. At times, there were over 200 raves happening in south London every week. Newspapers reported the phenomenon, and it was exploding in the UK. I personally think this was the last time a generation had such a cohesive, music-driven revolution. This outbreak in partying would be known as the Second Summer of Love because of its parallels to the 1969 Summer of Love. Instead of rock, it was electronic beats led by the Roland 808 and 909, and instead of LSD, it was MDMA. Like they say: Given enough time, history rhymes.

By 1991, the UK scene had sped up the tempo and feel of the tracks, while adding samplings of breakbeats from old disco and funk records (like James Brown) to form the rave and hardcore sounds. In the mid-'90s, UK garage (one of my favorite electronic genres till this day) emerged out of London.

The electronic sound kept evolving over time to yield drum and bass, trance, EDM, ambient and dub. Today there are countless forms of electronic dance music. It wouldn't be a stretch to say that most music today has some sort of electronic element.

A Wrap-Up of the '80s: The Fall of the USSR

Toward the end of the 1980s, things were looking up. The world economy was back on track, technology was making life easier, and kids had a new form of music that they were dancing to. The communist threat was also waning. The old Soviet leaders had died during the first part of the '80s, and the new head of the USSR was a reformer who understood the people. His name was Mikhail Gorbachev.

Under Gorbachev, the Soviets introduced the new policy of *glasnost* (openness) and a need for *perestroika* (economic restructuring). Censorship would be largely repealed, and citizens were free to question their government. It was a new way of doing things. However, some Eastern-bloc countries were slower than others to adopt the new policies. Hungary and Poland were the first to embrace liberalization, and slowly others (though not all) began to follow.

When I was in Germany recently, I was fascinated by the history of East and West Germany and how different life was between the two. In 1989, East Germans were taking part in protests. They had been locked in what amounted to a Soviet-controlled prison and wanted their freedom. In the winter of 1989, the Berlin Wall finally fell, after twenty-eight long years. Families that had been split apart were now reunited. I talked to a former East German local at a bar and he recalled being five years old when the wall fell. He told me it was the strangest experience having been told a lie your whole life, and then having it vanish overnight.

The Soviet Union had fallen, and the Cold War was over. Relief would give way to prosperity and technological expansion in 1990s, what some would call the last great decade.

PART 3

THE GREAT EXPANSION

CHAPTER 12

Digital Nostalgia
1990–1999

When the '90s rolled around, the overall mood was light. As we saw, the Cold War had ended, though the Gulf War would soon begin. Important figures from the decade included OJ Simpson, Princess Diana, and Monica Lewinsky. This is the first decade in this book that I actually lived in, so it's a special time for me.

Grunge, a new type of music and style of dress, emerged from Seattle, led by Nirvana. Grunge was angry, cynical, and the anthem of Generation X. Punk was thrust back into the spotlight by bands like Blink-182 and Green Day. A commercialized version of the '60s sound emerged out of the UK, from groups like Oasis and the Verve. Rave, hip-hop, and pop from groups like the Spice Girls were all sharing the limelight. Justin Timberlake, Christina Aguilera, and Britney Spears all got their start on the Mickey Mouse Club.

Sitcoms were king, and Seinfeld, "a show about nothing," became a worldwide sensation. In 1991, movies like *Titanic*, *The Lion King*, *Forrest Gump*, *Men in Black*, *Jurassic Park*, *Home Alone*, *The Matrix*, and *Toy Story* were released and have all remained classics.

Meanwhile, the internet was beginning to connect people—across the globe. Multiculturalism was on the rise, which was reflected in the popularity of "world" music at the time. In the same vein, there was also a keen interest in foreign films.

Fashion was diverse, with gothic styles, grunge flannels, baggy clothes, and a faint hippie tinge all being part of the times. Extreme sports and saying "radical, dude" were all the rage. I remember TV ads aimed at kids always had someone on a BMX bike or skateboarding, just to seem hip. I bought a cheap fluorescent yellow plastic BMX bike as an eight-year-old and it snapped in half when I slipped and fell in the pool while riding in the backyard. I cried.

Western society was becoming more liberal, and economies were thriving. This created a strange dichotomy of eastern philosophies (especially environmentalism) and entrepreneurship being embraced by the young. The good times of economic prosperity (in the '80s as well as the '90s) brought about an echo generation of the baby boom, known as the millennial generation. The leading edge of millennials would be the first generation to grow up in a connected world. Famous millennials include Beyoncé, Roger Federer, Mark Zuckerberg, Prince Harry, and Usain Bolt.

As the economies of most western countries began to grow, jobs were created, inflation decreased, and the explosion of the internet caused unprecedented prosperity. 1991 to 2001 was the longest economic expansion in history. It was so all-encompassing it was called a "new economy," ushered in by technology.

It seemed there would never be another recession again. In the US, hourly wages went up 10 percent in four years (1996–2000). That's like everyone getting a bonus.

The invention of the Web expanded the use of the computer. Instead of being strictly a tool for amplifying the mind, it could now be a tool for communication and knowledge discovery. In the '90s, email became the hottest thing ever. It was instant and cheaper than traditional mail (in fact, it was free). In addition to communication via email, chatrooms also rose to prominence. Virtual communities from around the world could gather and discuss topics of shared interest.

Technology hardware capabilities exploded. Between 1990 and 1999, PC power skyrocketed, going from an average of 1,048 KB to 1 GB of RAM. In contrast, average computer memory only went from 128 kb to 1,048 KB in the 1980s. Moore's Law was being proven.

A digital revolution would begin in the latter half of the decade with the introduction of the mp3 player, online download services, libraries turning paper into bytes, and the increasing availability of digital cameras.

Apple saw a massive slump in Mac sales from 1990. Microsoft Windows 3.0 (a graphical interface on top of DOS) sold 30 million copies within a year, pushing the Mac into a corner. IBM also tried to tackle the monster they had created in Bill Gates by releasing their own operating system, OS/2. It failed.

Science would also get a few breakthroughs, with cosmic dark energy being hypothesized to make up 70 percent of the energy universe, DNA being used in criminal law, Dolly the sheep being the first successfully cloned mammal, and the Hubble space telescope being launched, forever changing astronomy.

The First Text Message (1992)

Before the '90s, teenagers had no way of getting a mobile phone, they were too expensive. Payphones were the only way of communicating outside the home. You had to be where you said you'd be at a particular time, or else you wouldn't be seeing your friend that day. (This was fantastic for integrity, however!)

In the late '90s, mobile technology rapidly accelerated, and prices came down to the point where most young people (or their generous parents) could afford one.

Along with the boom of the mobile phone came text messaging, which brought communication to another level. It was great for quick relays of information: "I'm running 5 mins late," "I'm outside," "Where r u?," that kind of thing. Today trillions of texts are sent each day.

The actual name for what we call texting is Short Message Service (SMS). The "short" part comes from the maximum size of a text message, which is 160 characters.

The world's first SMS was sent by twenty-two-year-old Neil Papworth in the UK. Papworth was a developer for a telecom contractor tasked with developing a messaging service for Vodafone. This first text read "Merry Christmas" and was sent to Richard Jarvis, a director at Vodafone, who was presumably stuffing his face at the office Christmas party. Mobile phones didn't have keyboards at the time, so Papworth had to type the message on a PC.

So that was it, right? After this point, everyone could go about texting their buddies. Well…not so fast. Most early GSM mobile phone handsets didn't have the ability to send text messages. The first SMS's were provided in a one-way stream from network providers to customer handsets. These were network notifications—billing alerts and the like. It was the Finnish firm Nokia that first introduced user-sent text messages in 1993.

In 1999, texts could finally be exchanged between different networks. By 2007, there were more text messages sent than calls. Today, SMS is the most widely-used data application in the world, with 81 percent of mobile phone subscribers using it.

The First Smartphone (1994)

If you ask people what they think the very first all-touchscreen smartphone was, most would probably say the original iPhone, or if they knew a bit more about technological history, the LG Prada. They would be wrong.

As you recall, Martin Cooper shocked the world with the very first mobile phone, the Motorola DynaTAC x 8000, in 1984. Although the breakthrough device allowed walking and talking wirelessly, it was by no means "smart."

For the first smartphone, we have to fast-forward to 1992. This was a time when primitive personal digital assistants (PDAs) were becoming hot possessions, a time when laptops themselves were just becoming feasible. In 1992, while other companies were looking for the right combination of communication capabilities, IBM already had a prototype and unveiled it at a Comdex computer industry trade show.

History was very quietly made when IBM showed off the small prototype phone they had been developing. It included PDA features and a large 4.5-inch touchscreen. The phone was refined further and put on the market in 1994, with the name the Simon Personal Communicator.

So, what could this phone actually do? Well, in addition to its ability to make and receive phone calls, Simon was able to send and receive faxes and e-mails, and included several other apps, like an address book, calendar, appointment scheduler, calculator, world time clock, and a notepad, all through its touch-screen display and stylus. It had one megabyte of RAM, one megabyte of MB flash storage and a 16-megahertz processor, which was actually almost as fast as a low-end desktop computer of that time. It was so advanced that people didn't even know what they were looking at. The IBM Simon was short-lived, being discontinued less than a year later, in February 1995 due to the rise of thinner flip phones. It sold 50,000 units.

The IBM Simon doesn't sound like much today, but at the time it was an absolutely revolutionary concept. So revolutionary, it would take another thirteen years until the smartphone industry realized, once again, that a screen which adapted to your applications worked better than fixed buttons. It would take the iPhone in 2007 to popularize this concept again. We'll learn the amazing story behind the iPhone in the next chapter.

THE WEB EXPLODES (1994)

The internet is a large part of our lives. Countless technologies have been built to utilize it. There are over 1.8 billion websites on the web and 4 billion users today. But how did it rise to such prominence after its inception?

In 1989, as you remember, our buddy Tim Berners-Lee built the web in an office at CERN on a NeXTSTEP computer. The idea was to make navigating through information of *any* kind, easy. In 1991 CERN published the code for the World Wide Web, and within a few years, it was taking over the planet. It began in August 1991 with one website, *The World Wide Web Project*. By 1992, there were ten websites. Initially, there was no commerce allowed on the internet, but this was overturned by Congress in 1992. The following year, CERN made the web free and public. By that end of the year, there were already 14 million internet users worldwide, but only 130 websites.

However, at this time, the web was just lists of uniform text. It wasn't much to look at, and you had to memorize code to get around. Only nerds could use it. Web accessibility was thus limited, without an easy way to navigate or browse. This is when graphical browsers stepped in; once again, graphical interfaces did what they always do—make technology easy for anyone to use.

Released in 1993, Mark Andreesson's Mosaic browser was the first popular graphical web browser. With his technology, everyone could now easily navigate the web without a need for coding knowledge, although they were still unable to search like we can today. We will look into how this latter issue was solved later in the chapter.

Mosaic spread through the internet like a global wildfire. The web went from 130 websites (pages) in 1993, to 2,738 in 1994—that's over 2,000 percent growth. Mosaic would later become Netscape, and Netscape went from zero to 65 million users in just a year and a half. This was a record for any software at the time.

I have my own memories of Netscape. When I was around five or six years old, my dad was a university lecturer. He would let me sit in his office and use his computer after school, while he finished up at work.

I remember using Netscape Navigator to find all the information I could about space—from planets to theories on black holes. The internet was so easy to use that even a kid could do it.

Netscape made $75 million in 1994, and $370 million in 1995. They became the fastest growing company in history. But Bill Gates would have something to say about that.

After Netscape's success in 1995, Bill Gates told Microsoft staff that the internet was *the* priority. The result was Internet Explorer, which was free, while Netscape cost $49 per package (for businesses).

During the late 1990s, everything seemed to speed up tenfold. Time itself became compressed, now measured in "internet years" by those in the industry. There was no time to take a break. Competitors would rise and fall. For comparison, traditional company growth might be 10 percent a year during that time, while internet companies could grow 15 percent a *month!* By this stage, the PC was the single biggest wealth-creation product on the planet.

There was a labor shortage in programming. The industry in fact had zero unemployment. Two years of web experience in this time equated to ten years in the previous generation of programming. Due to a high level of English language competency, India saw huge interest from internet companies as they employed Indian locals. In fact, third-grade students were learning computing in Bangalore in 1997.

The Problem with the Early Web

Although the Netscape graphical web browsers made it easy to get around, a new problem arose. As the internet grew, and more content came online, it became harder to find exactly what you were looking for. If only there were a way of organizing it all...

Initially, online information was organized in a catalogue, like in a library. But this was an old way of thinking about things. Better attempts at organizing information like Yahoo! (more on them later) would manually categorize websites in a searchable index. Unfortunately, this took a long time, and sometimes what you were looking for couldn't be so easily categorized. A more exact way of arranging things was needed, a new way of thinking. Instead of categorizing information, why not make it a popularity contest? The link that gets the highest number of clicks rises to the top. Later in the chapter, we'll see how this idea led to Google.

The Rise of AOL (1994)

At the beginning of the explosion of the Web, a proliferation of Web companies arose to try and make sense of this new paradigm. One of the greatest of these early companies was AOL.

To begin their story, we have to travel back to 1985.

At the time, there were some little-known but curious efforts at creating a closed Web (before Tim Berners-Lee's World Wide Web). In 1985, an online service called Quantum Link (Q-Link for short) was established. It was based on hypertext, but was made strictly for Commodore 64 computers, and hence was only available to Commodore 64 users. It featured animated graphics, online chatrooms, online multiplayer games (such as Hangman), and learning services with research articles and resources. It even featured the first graphical large-scale online game.

In 1985, Q-link had 100 users; in 1986, this became 50,000 users. It was clear that connecting computers through HyperText was a powerful thing and people wanted it, even at $22 per month. Q-Link was thought to be the next big thing, though it was only open on weekday evenings and on weekends.

In October 1989, after a failed partnership with Apple, Q-Link changed its service's name to America Online. AOL morphed into a user-friendly way of getting people online.

In 1991, AOL for DOS and Windows would be launched, and the company saw incredible growth from this new user base.

Jan Brandt, head of marketing, was probably the person to thank for AOL's incredible growth. In 1993, she did something that went against all business sense of the era.

Jan Brandt, in her previous occupation, had sent out free copies of her book in the mail, and the extra awareness translated to sales. This marketing technique had been extremely effective. So, when AOL brought her on board, they decided to try it. They sent out approximately 90,000 AOL floppy disks with free software and a trial membership. They expected only 1 percent of floppy disk receivers to sign up, but 10 percent did, and this set the company off. (AOL later did this again with CDs. At one point, 50 percent of the CDs produced worldwide had an AOL logo!) AOL quickly became the number-one online service, thanks to this strategy. This marketing idea of Jan Brandt's would be known as "carpet bombing."

In 1993, the opportunistic Bill Gates saw an opportunity in AOL and decided to meet with the team. The factions didn't see eye to eye, so Microsoft decided to work on their own web projects, which evolved into MSN Messenger and Internet Explorer.

Over the next several years, AOL partnered with educational bodies such as the American Federation of Teachers, National Geographic, the Library of Congress, and the US Department of Education to offer the first real-time homework help service, the first online courses, and the first online exhibit.

Within three years, AOL's user base had grown to 10 million people.

In 1996, Microsoft and AOL made an agreement to bundle AOL with Windows software. In 1997, AOL experienced immense growth, with more than 34 million subscribers. In November 1998, AOL announced it would acquire Netscape for $4.2 billion. They would also release AIM, AOL Instant Messenger. By this time, a new person was signing up every six minutes. In January 2000, AOL and Time Warner announced plans to merge, forming AOL Time Warner, Inc. It would be the biggest media company on the planet. AOL got so big, it was often mistaken for the internet itself.

Where Did It All Go Wrong?

No one talks about AOL today, so something must have gone wrong at some point. What happened?

The problem begins here: The AOL portal allowed you to browse categories, which was fine before search engines came on the scene, but this wouldn't last. Also, when the web exploded in popularity around 1994, there was a sudden availability of free independent websites. This broke the business model that had supported the rise of AOL in the first place.

However, the real blow came with the dot-com crash in 2000. AOL suffered the biggest corporate loss in history, at $100 billion. By this stage, people were happy to use a search engine like Google to get what they needed from the web. There was simply no use for portals that arranged the web into categories. In other words, there was no more need for AOL and others like it.

PLAYSTATION (1994)

Who can dislike Sony's PlayStation? It was so ahead of its time, bringing countless hours of fun and ushering in the 3D era of gaming. But where did it all come from?

Despite its place in the heart of gamers, the PlayStation was a product that originated out of spite. Let's take a look at this story.

The PlayStation started with one man: Ken Kutaragi, a Sony engineer. He first became interested in working with video games in 1988, after seeing his daughter play games on the NES. Kutaragi approached Nintendo to create a CD-ROM drive for the upcoming Super Nintendo System, the Super Disk. Nintendo agreed, and Kutaragi brought together a research and development team within Sony, in 1990. The following year, at CES (the Consumer Electronics Show), Sony would show off their SNES-CD, now code-named the PlayStation. It was a multi-media powerhouse, capable of playing not only Nintendo games, but also audio CDs, data CDs, and video CDs—a pretty big deal for 1991.

Unfortunately for Sony, disaster was around the corner. Just one day later, Nintendo would surprise everyone with an announcement: they announced that they would break up their partnership with Sony and begin a partnership with Philips instead. The SNES-CD would never see the light of day.

This made Sony president Norio Ohga furious. In an attempt to ease the situation, Nintendo told Sony that they could have a non-gaming role in the new partnership. Kutaragi rejected the offer and went to work on a rival system. Unfortunately for Kutaragi, other Sony employees weren't happy with the idea of fully committing to the video game industry. Kutaragi and his newly formed team had to move to the Sony Music headquarters in another building.

In 1993, Sony Computer Entertainment was formed by the CEO of Sony Music and Kutaragi. The involvement of Sony Music was actually a blessing in disguise, as now, marketing, creation, and the manufacturing of discs could be done in-house, saving both time and money.

The following year, in a Tokyo hotel, Sony showcased technical gaming demos and revealed their distribution plan to software companies. The session opened the floodgates for keen developers.

Both Electronic Arts and Namco were interested. The main selling points of the system (soon to be PlayStation) were 3D graphics and CD capabilities, a major step above the cartridges of the time. With a 34-megahertz CPU and MB2mb of RAM, it was also powerful for its day.

In December of 1994, the Sony PlayStation was let loose in Japan. Its main competition was the Sega Saturn, though the PlayStation won out. The lower price point helped Sony, but in addition to this, the Saturn's dual-core CPU made program development hard, resulting in fewer games. A world release of the PlayStation followed in 1995, and popularity soared. The rest is history.

The PlayStation was the first console to exceed 100 million sales and went on to become one of the iconic consumer products of the '90s. Not even the Nintendo 64 could dampen the success of this Sony machine. The PlayStation's successor, the cleverly titled PlayStation 2, would later become the best-selling home console of all time, selling 155 million units. PlayStations 3 and 4 would also both be smash hits, establishing Sony as a leader in living-room entertainment.

Xs and Os

When designing the controller, Sony designer Teiyu Goto decided to use four symbols. The triangle represents the player's point of view, the square represents a piece of paper (for menus or other in-game documents), and the O and X buttons respectively represent "yes/ right" or "no/wrong" options—this was reversed in western territories, so that the X button became the de facto "move ahead/ action" option.

The first major Sony-developed game was *Motor Toon Grand Prix*, by a small team called Poly's. This team would go on to become Polyphony Digital, creator of the legendary Gran Turismo series.

In 1994, Namco also developed a test board, "System 11," to simulate the PlayStation 1 during development. The first major title to use the board was a test program for a fighting game experimenting with texture-mapped 3D characters. The experiment would eventually become *Tekken*, a key PlayStation hit.

YAHOO! (1994)

Remember Yahoo!? When the online world was still taking shape, this company was one of the early pillars of the web's construction.

Yahoo.com was founded in 1994 by two Stanford graduates, Jerry Yang and David Filo. Yahoo! was to act as a portal to the internet at a time when it was still a mess of unorganized websites and information. At the time, there were few meaningful ways to quickly browse for what you wanted. The company went public in 1996, and by 1997 it had the second-highest traffic on the internet, after AOL.

In 1998, Yahoo! was approached by two young Stanford PhD graduate students, Larry Page and Sergey Brin, who had a new product. The pair had just created PageRank—a search engine algorithm and a quick way to get relevant information on the internet. The Stanford students were asking for $1 million. Yahoo! said no, as they thought PageRank would take users away from Yahoo!'s website—decreasing its traffic and ad revenues. PageRank would eventually become Google.

The dot-com crash would hit all tech companies hard, but Yahoo!, Google, and a few others survived the storm. Yahoo! didn't escape without some wounds, though. Their stock price dropped from over $118 per share to just $4.06. The company would never see such a high valuation again.

Yahoo! was the most visited site on the internet in 2002, but after the dot-com crash, Web portals (the old way of organizing information) became irrelevant, as Web users turned to search engines like Google to find what they wanted. Yahoo!'s technology just couldn't keep up. In an attempt to stay afloat, Yahoo! purchased 114 companies over the course of its lifetime. Despite this effort, none of the purchases turned into a hit that could save Yahoo!.

Yahoo! eventually just got left behind, in all areas of business, by Web-based companies that did it better. Yahoo! mail got overtaken by Gmail and Hotmail. Yahoo! messenger got enveloped by WhatsApp, WeChat, and Facebook Messenger. Yahoo! News also got swept over. People went to social media to learn about current events on Facebook, Twitter, dedicated news websites, or even YouTube.

As users drifted away from Yahoo! services, so did the ad revenue and hence, company profits.

WINDOWS 95 (1995)

By 1995, Windows was no longer a hack-job plastered on top of DOS. Previous versions of Microsoft's operating systems had been notoriously unstable. They frequently crashed without a clear reason and were complicated to set up. Windows 95 would replace DOS to become a stable and reliable breath of fresh air from Microsoft. This version of Windows introduced numerous functions and features that remained for decades to come. Such Windows 95 introductions included the "Taskbar" and the "Start" button. Interestingly, Brian Eno designed the start-up sound, a piece of audio giving instant nostalgia to anyone who listens today.

PIXAR ALMOST FAILS, BUT *TOY STORY* SAVES THEM (1995)

Pixar began in 1979 as the Graphics Group subsidiary of the Lucasfilm computer division—this was the graphics group headed by Edwin Catmull (the guy who scanned his hand and made a 3D film in the 1970s). The Graphics Group had wanted to create a computer-animated feature film from its inception. In 1979, they attempted to produce such a film for an early 1980s release. The project was called *The Works* and it was to be ninety minutes long. But it would never be.

The computers of the time were still too slow and underpowered to create a feature-length film. During the process, one of the team members calculated the required time to output the number of rendered frames needed for the film: It would take seven years. The film was abandoned in 1986. *The Works* would have been the first entirely 3D CGI (computer-generated imagery) film, had it been finished as intended.

By the time it was abandoned, just over a tenth of the film had been completed. To fix the problem of slow computers, the Pixar team decided to work on their own computer hardware. The result was the Image Pixar Computer, at the nice low price of $219,000. It was powerful for the time, around fifty times faster than conventional equipment. A scene that would traditionally take fifteen minutes to render (draw and create) now only took eighteen seconds.

In 1986, Lucasfilm was looking to sell the graphics department (Graphics Group). A few months later, Graphics Group would spin off from Lucasfilm as Pixar, a standalone corporation, with Catmull as president. As exciting as this was, the fledgling Pixar needed investment money. After over forty rejections from investment companies, Pixar approached Steve Jobs. Jobs (now CEO of NeXT, after leaving Apple), liked the idea of combining art and technology, and invested $10 million in the new company. Via this transaction, Jobs became the majority shareholder, acquiring 70 percent ownership. Despite the change in ownership dynamics, Pixar's ultimate goal was still the same: to create a fully computer-generated feature film.

Walt Disney Studios subsequently bought a few Pixar Image Computers, and custom software written by Pixar, to be part of their Computer Animation Production System (CAPS). With CAPS, the painting and inking of 2D animation was now computer-assisted—saving effort and money.

The first film to use CAPS was *The Rescuers Down Under* (1990). It was the first feature film to be 100 percent assisted by computers. *Aladdin* (1992) and *The Lion King* (1994) also used the system, which would be in use until *Pocahontas* in 1995.

With the limited clientele of Disney for CAPS, computer sales were lackluster for Pixar. Realizing financial trouble was coming, Jobs suggested releasing the Pixar Image Computer to other markets that would be interested: engineers, geologists, and the medical field. Pixar employee John Lasseter went to work creating short, completely CGI demonstrations to show off the technology. These demos included *Luxo Jr.*(1986) and *Tin Toy* (1988). Lasseter—who had been obsessed with toys from childhood—had a knack for injecting inanimate objects with personality and character, without them ever talking. This was the critical essence of early Pixar films.

The short CGI-animated films caused Disney to realize the full potential in this kind of filmmaking. In an ironic twist, John Lasseter had originally been fired from Disney for pitching a fully CGI film (after being shown *Tron* by two work colleagues). After seeing *Luxo Jr*—featuring the Pixar lamp—and other short CGI films, they now wanted Lasseter back. Defiant, Lasseter chose to stay at Pixar. He would tell Catmull, "I can go to Disney and be a director, or I can stay here and make history."

There's no difficulty guessing who John had been spending a lot of time with...*cough* Steve Jobs.

Despite ground-breaking demonstration films, the Pixar Image Computer never sold well. In fact, only three hundred units were ever sold. The company tried making CGI television commercials and selling software for Windows and Macintosh, but it wasn't enough. The inadequate sales threatened to put Pixar itself out of business, as financial losses grew. Jobs poured more money into the project in exchange for an increased stake in the company. His total investment was approaching $50 million. This gave him control of the entire company, but almost drove him into personal bankruptcy. Pixar was still struggling, and eventually had to sell their computer hardware business and lay off staff. Luckily, by 1991, Moore's Law (chapter 8) meant that animation-grade computers were finally *just* powerful enough to create and render a complete animated feature film (not just small animation demos) in a reasonable time.

Disney and Pixar Join Forces

Disney, who now realized they couldn't get Lasseter back, decided to strike a deal with Pixar instead. It was a historic $26-million deal to produce three computer-animated feature films, the first of which was *Toy Story*. Without this contract, Pixar probably would have closed its doors a few years later. Pixar got to work on *Toy Story* with Lasseter as the director and Jobs as the executive producer. The first 30-second test would be delivered to Disney in 1992, and the initial preview impressed them.

All did not remain well, though. The collaboration between the two companies became tense, as Disney executives continuously scrapped script ideas brought forward by Pixar. For example, Disney cast Woody as an edgy, tyrannical "jerk" (as Tim Allen would remark while recording the voice of the character).

At the halfway point of production in 1993, with heads and ideas clashing, a preview screening of *Toy Story* was shown to upper management at Disney. It was a disaster! Upper management absolutely hated it; indeed, so did Lasseter. As he put it, the movie was "a story filled with the most unhappy, mean characters I've ever seen."

The movie's production was shut down immediately, and it seemed like the dream of a CGI feature film would never come to pass.

But Lasseter didn't give up. He told Disney he would revise the script in two weeks. He somehow delivered, and the result was good enough for production to resume in 1994. For context, it's important to note that the production of such a film using early '90s technology was no easy task. Each frame took between forty-five minutes and thirty hours to render, depending on the complexity and detail in the scenes. Every eight seconds of film in which a character was speaking took the animators a week to complete. Every shot would pass through eight teams, and three hundred computer processors (running twenty-four hours a day) would render the final product.

But disaster was still looming. Despite receiving funding, Pixar continued to lose money: There wasn't enough income from the Pixar Image Computers, software, and CGI commercials to offset the cost of the *Toy Story* project. Jobs, as chairman of the board and now full owner, was getting nervous. Who wouldn't be after spending $50 million of their own money on a seemingly failing project? Even as late as 1994, Jobs contemplated selling Pixar to Hallmark Cards, and also to Microsoft co-founder Paul Allen.

This sentiment would change as the film neared completion. As the beautiful computer-generated world of Toy *Story* began to take shape, Jobs marveled at what he saw, and realized that Pixar would be historic after all.

Toy Story went on to gross more than $373 million worldwide. When Pixar held its initial public offering on November 29, 1995, it exceeded Netscape's as the biggest IPO of the year.

I'm sure that, when you saw it for the first time, you were just as blown away as I was. The film looked more realistic than any computer-generated imagery I had ever seen. Every leaf, every small detail, was spot-on. It captured my imagination and the imagination of a whole industry. *Toy Story*-inspired material included toys, video games, theme park attractions, merchandise, and two sequels. *Toy Story* made toys cool again for the '90s, and there was no doubt that *Toy Story* single-handedly saved Pixar.

Without global positioning systems (GPS), you wouldn't have Google Maps, Amazon delivery, Lyft, Uber, or many other things.

It all started back in 1973, when the GPS project was launched by the United States government. At the time, navigation systems were fairly poor. A more accurate system was needed. During the space race in the 1960s, classified studies were being conducted on satellite technologies.

The GPS system was originally for US military purposes, but in 1983, a plane crash would change that. A Korean Airlines flight was shot down after straying into restricted USSR airspace. Two hundred and sixty-nine people lost their lives. President Reagan's team theorized that the incident wouldn't have happened if the aircraft had had a more accurate navigation system. GPS fit the bill perfectly, so Reagan declared that GPS should be freely available for civilian use, for the common good of mankind. The Global Positioning System became fully functional in 1995, after the launch of the twenty-fourth satellite in the GPS system. The program is said to have cost $5 billion.

Today we have GPS technology with an accuracy of up to 3.5 meters: astounding when you think about it. Currently there are thirty-one GPS satellites in orbit.

AMAZON KICKS OFF E-COMMERCE (1995)

A decade ago, the idea of browsing online for almost any item you wanted, and having it delivered to your door within the same hour, without even leaving your chair, would have sounded like science fiction. But this same-day delivery is what Amazon does in Europe and the United States on a daily basis. Today, Amazon Global hosts the sales of 350 million products in 16 countries and delivers to 185 countries, 10 countries short of the entire world. How did such a massive company start?

In an echo of Apple's founding, Amazon was started in a garage, this time by Jeff Bezos. The idea first came to him when he was working at a Wall Street firm, D. E. Shaw & Co., as a vice president. In his work Bezos was noticing the rapid increase in internet users and saw potential. Now that people were connected, it was only a matter of time before someone created an online store.

During discussions between Bezos and his boss at the firm, the concept of an online store came up, and the idea for Amazon was born. Bezos ran with it. He concluded at first that an online store that sold everything would be impractical, so he made a list of twenty possible product categories, including computer software, books, office supplies, apparel, and music. The option that seemed to fit best for a first product was books. Books were the same, no matter where you bought them, and offered diversity in terms of choice.

Together with his wife, MacKenzie Tuttle, and start-up veteran Shel Kaphan, Bezos created the new company. He first registered the company in July 1994 in Washington State as Cadabra, Inc. The name was misheard as Cadaver. A company name that could be misheard as a dead body wasn't ideal, so they chose a new one. Bezos says they chose Amazon because it is the largest river, mirroring his goal of becoming the largest online retailer. Sites on the web were also listed alphabetically back then, and he wanted his site to be among the first web pages shown. Amazon.com was registered on November 1, 1994, and the team began to grow.

In a test run, the first book was sold on April 3, 1995. The book was *Fluid Concepts & Creative Analogies: Computer Models of Fundamental Mechanisms of Thought* by Douglas Hofstadter. Thrilling reading! On July 16, 1995, the site went live and became visible to all internet users. It opened with a searchable database of over one million titles, even though it only had around 2,000 in stock at its Seattle warehouse. The rest would be ordered as needed from wholesalers.

There was no one assigned to pack the books at Amazon at the time, so the team would take orders during the day and pack the books during the night. The books were driven to the post office the next morning. This humble arrangement changed when Amazon was featured on the home page of the Yahoo! portal. After this exposure, Amazon was swamped with orders. Within the first month, they had shipped books to forty-five countries (at the time, an American address wasn't needed).

Bezos predicted $74 million in sales by 2000, if things went moderately well, and $144 million if they went extremely well. In fact, sales in 2000 amounted to $1.64 billion. This was the start of Amazon's frantic expansion. Amazon's motto at the time was "Get Big Fast."

On May 15, 1997, Amazon held its Initial Public Offering (IPO), raising $54 million and valuing the company at $438 million.

Having predicted that the small size of the reading community would be a bottleneck to their growth, they decided to introduce new products. DVDs and music were chosen. They opened a new distribution center in Delaware and expanded their existing one by 70 percent.

Between 1998 and 2000, Amazon acquired IMDb, Telebuch, and Bookpages, which gave them strategic access to customers in the UK and German markets. During this time, they also expanded their product line to offer toys and electronics.

Amazon was growing rapidly, and the world took notice. Bezos was named "Person of The Year" by *Time* magazine. Surprisingly, Amazon did not post any profit until the fourth quarter of 2001, when it made $5 million in profit (despite the incredibly high sales figures, they were spending all their money on growth). Bezos said this was because he emphasized customer experience over profits. Amazon continued to grow and became a dominant force in the 2010s (chapter 14).

MP3 PLAYER (1997)

The Visionaries

Carrying vast libraries of digital music in our pockets is now an everyday occurrence. If we try and think back to when it all started, the original iPod comes to mind, along with other early mp3 devices. But, surprisingly, the history of the MP3 player has its roots all the way back in the late '70s.

In 1979, British scientist Kane Kramer and his friend James Campbell came up with the idea for a cigarette-pack-sized, portable, solid-state (chip memory) music player. Kramer was twenty-three, Campbell twenty-one. The system they envisioned, called "IXI," had a display screen and buttons for four-way navigation. Is this sounding familiar?

Solid-state storage at the time was severely limited, but Kramer fully expected this to improve. He foresaw a market for reliable, high-quality digital music players that would be popular with both consumers and record labels. Kramer described methods of digital distribution (similar to the iTunes store) and music held on swappable chips. His vision embodied the end of physical media in 1979!

Kramer filed a patent to start his own company. Things were going well initially: there was a projection of $328,000,000 worth of orders, and he even had Sir Paul McCartney as a financial backer.

In 1988, the patent was at risk of lapsing, and Kramer's company needed to raise $320,000 to renew it. There was a boardroom dispute on the issue, and the patent lapsed, meaning Kramer wouldn't get recognition for any products using this idea. As a consequence, this early British MP3 player was never put into full production. Besides this, it would still be four years before the mp3 (compressed audio) format came into existence.

The First Production MP3 Player (MPMan)

The first portable MP3 player was the MPMan (a play on Walkman, I assume), launched in 1997 by Saehan Information Systems, a spin-off from Samsung. The Flash-based mp3 player offered 32 MB and 64 MB (six or twelve songs) storage-capacity options. Also included was an LCD screen to tell the user what song was currently playing.

Music has always been very important to me, so I distinctly remember my first MP3 player in 2004. It was a tiny silver Conia 1 GB device, about the size of a thumb. People at my school couldn't believe something so small could hold a whole gigabyte. I remember classmates repeatedly asking to check the storage capacity on the side to make sure I wasn't lying.

What Happened to Ken Kramer?

In 2008, when Apple acknowledged Kramer as the inventor of the iPod technology, he was a struggling furniture salesman. He explained the events in an interview with the Daily Mail:

> *"I was up a ladder painting when I got the call from a lady with an American accent from Apple, saying she was the head of legal affairs, and that they wanted to acknowledge the work that I had done. I must admit that at first, I thought it was a wind-up by friends. But we spoke for some time, with me still up this ladder."*

The revelations came to light after patent-holding company Burst sued Apple, claiming the iPod infringed on its patents. Apple flew Kramer to the US to give evidence in its defense and used his original 1979 drawings of the IXI as evidence that Kramer, in fact, was the iPod's inventor. Apple essentially used Kramer to get out of legal trouble, but he did gain something for his time.

After the ordeal, he explained: "Apple did give me [an iPod], but it broke down after eight months."

We'll catch up with the iPod in the next chapter and see how 1,000 songs in your pocket started a revolution.

STEVE JOBS RETURNS TO APPLE (1997)

Say what you want about the man, but there's no doubt that Apple needed Steve Jobs. This is how Apple was doing in the 1990s while Steve was away on his wilderness experience:

1991: Apple unveils a portable Mac called the PowerBook; it doesn't do well.

1993: Apple introduces the Newton, a hand-held, pen-based computer. It is also a flop.

The company reports a quarterly loss of $188 million in July. John Sculley is replaced as CEO by Apple president Michael Spindler. Apple restructures, and Sculley resigns as chairman.

Meanwhile, at NeXT Computer, Steve Jobs decides to focus on software instead of computer hardware.

1995: Microsoft releases Windows 95, which is easier to use than previous versions and is similar to the Mac system. Apple struggles with competition, component shortages, and errors in predicting customer demand.

While this is happening, Pixar's *Toy Story*, the first commercial computer-animated feature, with Jobs as executive producer, hits theatres. Pixar goes to Wall Street with an IPO that raises $140 million.

1996: Apple announces plans to buy NeXT for $430 million. In the acquisition deal, Jobs is to be appointed an adviser to Apple. To me, it looks like Apple was extending an olive branch, realizing its mistake of letting Jobs go.

As mentioned in the previous chapter, NeXTSTEP became MacOS X (MacOS ten) and the basis for iOS in the iPhone. More on this in the next chapter.

1997: After the NeXT purchase, Jobs gets a seat on the board. From here, he manages to climb his way back to the top of Apple.

The famous advertising slogan "Think Different" is launched in this year.

1998: Apple returns to profitability and shakes up the personal computer industry with the candy-colored, all-in-one iMac desktop. Apple discontinues the Newton and regains its vision.

The Founding of Google (1998)

While Apple was getting back on its feet, computers were appearing inside the home in ever-greater numbers. Unlike the 1980s, this was different. Now the average person could reach out and instantly access the rest of the world through the Web.

An endless catalogue of knowledge and information was being built all over the Web, but it would all be useless if it was almost impossible to find what you wanted. The Web had to be organized.

In 1995, Larry Page was working on the Stanford Digital Library Project. One night, he had a dream. In that dream, he imagined that he could store the whole Web on old university computers. After Page awoke, he did some calculations and theorized that it was possible—but only if you stored the *links* to the websites, not the actual websites themselves. Inspired, he got to work building a program that could do this.

Meanwhile, Sergey Brin, a Soviet immigrant, was studying computer science at Stanford and had an interest in data mining and trends. Page met Brin in 1995. Larry Page let Brin know about his dream of downloading the entire web. Due to his interest in data mining and trends, the web seemed like the most interesting data set possible for Brin.

Page's father was a computer scientist, so he had grown up with computers around the house. Page was playing with his first computer in 1978, when he was six; this situation was unique for the time. According to Page, he was the first kid in his elementary school to turn in a word-processed piece of homework. As a kid, he would read popular science and computer magazines and was interested in the way things worked.

Brin was also inspired by computers, especially the power they give us. He was interested in leveraging that power to do useful things beyond video games. In the '80s, he and a friend would play on an Apple Mac by programming primitive AI, a gravity simulation, a chat bot (conversation simulator), and a program for recognizing letters. That experience would be pivotal in Brin's life.

Sharing similar childhood experiences, Page and Brin worked together to organize the world's information. To tackle this problem, Page began by reversing the links—starting local with the Stanford homepage. He found there were 10,000 external links trailing back to the Stanford homepage. There was a problem, though: Larry Page could only show ten link results on a screen. It was clear they needed a way to rank the linked pages so that the most relevant links showed first—kind of like a popularity contest.

After some tinkering, the results were what you'd expect: the pair successfully configured the Stanford homepage to be the most relevant result. It was actually Sergey Brin who had the idea that information on the web could be ordered in a hierarchy by "link popularity": a page is ranked higher if there are more links to it. Page and Brin didn't initially set out to start a search engine; their tinkering was pure curiosity and research, but a search engine came out of their efforts regardless. When they analyzed the data, the pair figured out they had the technology to enable a better way of *searching* for information.

They called their technology PageRank. It was named after Larry Page, but cleverly had a double meaning for its function of ranking web pages.

Page and Brin realized that organizing the world's information could be extremely impactful. Their product could truly be for everyone looking for information. Now they just needed a company to market it. They started a web company and called it BackRub.

BackRub began indexing (storing ready-to-be-ranked) websites in March 1996. From here, the index quickly grew. By August 1996, some 75 million pages had been indexed, which MB in total took up 207 GB of memory.

In 1998, they would patent the PageRank idea, and change their name from BackRub to Google, Inc.

As early as 2000, Page was talking about an AI taking all the web's information and providing an answer to any question. If you think about it, that's how we use Google today. When you're having a debate with a friend, or are just curious about something, what do you do? Google it!

In 1998, Google had 10,000 searches a day. In 2000 it had 50 million, and in 2018, it had over 3.5 billion.

Today, Google does a lot more than just search, with Gmail, Maps, YouTube, Android, data analytics, quantum computing, and AI being just a few of their twenty-first-century ventures.

Napster (1999)

As Google was revolutionizing search on the web, the world of music was also being transformed with the sharing of mp3 music. Today it seems like the current generation expects music for free. This mentality all started with Napster.

In 1998, Northeastern University freshman Shaun Fanning was just nineteen when he and a friend created Napster. It was a revolutionary concept: sharing stored music across the internet for free.

Fanning dropped out of college, and soon had Napster online and available to everyone. Word quickly spread through chat rooms, and there were 40,000 users in a few months.

Napster was the first large-scale peer-to-peer file-sharing software. Users could download files stored on other users' hard drives, as long as they were connected. Some early skeptics thought that nobody would be willing to open up their hard drives to the world—but for the reward of free music, it was a risk millions were willing to take.

Though CDs were booming in the '90s, they cost around $18 each, which was a high price to pay if you only liked one song on an album. By 1999, the music industry was making $15 billion in revenue in USA alone. Napster destroyed that in a flash.

Consumers could get the music they wanted for free. The file-sharing software was a smash hit on college campuses. By the year 2001, over 25 million people were using the Napster service.

Record companies were having none of it, and they called what Napster was doing straight-up stealing. In July 2001, the service had to shut down after a year-long legal battle.

But the idea had already permeated the collective consciousness. The cat was out of the bag, and free file-sharing sites began to spring up everywhere. Hypothetically speaking, I may or may not have used LimeWire, eDonkey2000, or Soulseek back in the day...

More lawsuits followed, but the damage was done. CD sales continued to plummet throughout the decade. It was Steve Jobs who would launch the iTunes digital music store in 2001, taking the idea of downloading any song you wanted legal.

This broke the stranglehold of the record labels. Later on, streaming services like Spotify would be the final nail in the coffin.

Emojis (1999)

Emojis were invented by Shigetaka Kurita in Japan, 1999. He got the idea from observing the relationship young people had with their pagers (remember those?). In the mid-'90s, Japanese youth loved a particular brand of pager, not only because of its functionality, but because it had a heart symbol that could be sent over text.

When a new version of the pager came out without the heart, there was an outcry, and many users left in search of another pager that provided the heart symbol. This inspired Kurita. In 1999, as he was helping implement a mobile internet system, he thought support for a selection of symbols for phones would be popular in Japan. He was right—about Japan, and the rest of the world.

The Y2K Bug (1999)

By New Year's Eve 1999, people were only concerned with two things: how hard they could party, and the Y2K problem.

The problem was that the two-digit date-storage system was going to roll over and read the last two digits of 2000 ("00") as the year 1900. It was suggested that it would throw the world into chaos. Nobody knew for certain how this rollover would affect important systems, but few were willing to take chances. In the late 1990s, there was a scramble for businesses, governments, and individuals to achieve Y2K compliance, spending billions of dollars in the process.

We may laugh at the Y2K issue now, but looking back on some of the predictions at the time, it was pretty scary.

The estimated worldwide cost of fixing the Y2K bug was predicted to be $1.1 trillion. It was thought everything from banks to nuclear facilities were at risk of catastrophic and unpredictable behavior.

The public were scared too. One American survey found that 42 percent of people planned to stockpile food and water, 34 percent expected banking and accounting systems to fail, and 51 percent planned to avoid air travel on or around January 1, 2000.

As midnight, December 31, 1999, approached, the world held its breath and...pretty much nothing happened. Except a world-wide hangover.

But, you may be surprised to know that some systems did fail. For example, a US military satellite surveillance system went down and wasn't recovered for several weeks.

By 1999, there were very few wars or geopolitical conflicts between the 6 billion humans currently on earth. There was peace and economic prosperity for most. The internet had transformed the world in nine short years. And mankind was at a point of no return. By 2000, we had entered what could be considered a post-modern era of living.

Technology Becomes Personal
2000–2009

Once the collective partying of the world was over and fears of the Y2K disaster had faded, it was all smiles for the year 2000. The twenty-first century had finally arrived.

The euphoria was short-lived, as the events of September 11 would shake the fabric of society. It was traumatic not only to Americans, but also to the rest of the world. I remember the next day, our school (in Western Australia) had students taken to the oval (the sports field) as a precaution. There on that oval, and in classrooms over the next few days, our teachers sat us down and we discussed our thoughts and feelings about the events. I remember a feeling of overwhelming disbelief.

As the decade progressed, manufacturing would leave the United States and Japan and move to other countries. The Euro—originally strictly a monetary agreement—currency would be put into circulation in 2002. China would see incredible growth, making up almost 9 percent of the world economy by 2009, at which stage it was the third largest economy.

In 2001, the Enron scandal and bankruptcy shook the business world. Similar criminal cases of fraud occurred in 2008, when hedge funds and banks passed off junk mortgage debts as premium assets. Meanwhile, Bernie Madoff took advantage of the desire to make a quick buck during the real estate bubble and created the world's biggest Ponzi scheme. Madoff *made off* with over $60 billion.

Reality TV became popular with *Big Brother* and *American Idol*—we had the Australian versions of both. I thought it was an innovative style of TV, especially the concept of *Australian Idol*. Shows like *Big Brother*, however, didn't seem to really be about much. I assumed TV shows of that type would be a fad. How wrong I was.

In the 2000s, R&B became popular again, as did emo. Groups like My Chemical Romance and Fallout Boy spread the emo genre in the early 2000s through sites like Myspace.

Fashion looked to the future early in the decade, with silver jeans not being unusual (I even owned a pair), frosted tips, skinny jeans, flares, and cargo pants. It was an eclectic mess.

The generation of this decade was Generation Z, mainly the children of Generation X. Gen Z was the first generation to grow up with social media like Facebook and YouTube. Due to globalization, they were generally more tolerant. They also had an entrepreneurial streak, especially after the great recession of 2008. In response to the recession, the alias Satoshi Nakamoto released the Bitcoin whitepaper in 2008 as an alternative financial system.

MSN Messenger made its mark in the 2000s. It was the coolest way to communicate. The hearts of teenagers would skip every time they saw their crush sign on. Windows XP, Vista, and 7 all debuted during the decade. In 2003, the 3G mobile network standard made mobile internet practical. Gmail launched in 2004.

By the middle of the decade, mobile phones became cheap enough that everyone had one. The Microsoft Zune and PlayStation portable were also hot items. This was an exciting period for consumer technology: things were always changing, and it seemed like it was all going somewhere.

Dot-Com Crash (2000)

The dot-com bubble is something that most people are familiar with: Young nerds making millions for any idea involving the internet. Well, not exactly, but it seemed like it. It all started in 1993, when the Mosaic web browser (later Netscape) made the Web a breeze to browse. Then, thanks to Google, anyone could search and find whatever they wanted. This opened the door to a huge market. By the late '90s, the internet was seen as a new economy: a disruptive magical concept that could transform the planet.

Yes, the internet did have the ability to change the world, obviously. But expectations grew too high too fast, becoming overhyped, and leading to fever-pitch levels of investment. People were throwing money at any company with a ".com" at the end of its name—even if they didn't have a real plan. Employees of these new companies would become instant millionaires as soon as their IPO opened.

Who needs a business plan, right? These star companies would throw lavish dot-com parties without batting an eyelid. It seemed too good to be true... Hold that thought.

Too Good to be True: The Unravelling

When the stock prices of these dot-com companies went up, people on the sidelines wanted to get in on the action. Soon, your local taxi driver was giving advice on which dot-com stocks to pick. It was time to get out—everyone who understood that these companies weren't making money did just that.

In the end, there were too many failed promises.

Investors eventually caught on, and the party came to an end in the latter half of 2000. The majority of the stocks were sold in a panic as everybody tried to rush out the door. By November 2000, over $1.7 trillion (75 percent of the value of internet stocks) had been wiped out. By 2002, $5 trillion had been obliterated from the entire stock market, as the contagion spread beyond tech stocks. Most tech companies got hurt badly in the crash. To give you an example of the roller-coaster ride, Blucora, a search engine company, was worth $1,305 a share in March 2000. By 2002, that price had fallen to $2 a share. Only the Googles and Amazons of the day survived. Since the crash was mainly due to tech stocks, the overall market picked up again in the coming year.

BLOCKBUSTER TURNS DOWN NETFLIX (2000)

From the mid-1980s to the late '90s, VHS was king. There was no other way to watch the movies you loved at home without being bound to the chains of TV programming. The problem was that VHS tapes could be expensive. For this reason, video rental stores like Blockbuster Video came in to fill that gap. They were the perfect solution and became a regular part of weekend plans for hundreds of millions around the globe.

Eventually, online video streaming services like Netflix, Hulu, and Amazon Prime Video destroyed the old video rental business model. Blockbuster came to the party late, even though it got an early invite.

In 2000, Netflix proposed that it would handle Blockbuster's online component, and in return, Blockbuster could host Netflix as an in-store component (that is, customers could pick up their DVD orders under the Netflix brand, thus eliminating the need for mailed DVDs, which was Netflix's business model at the time). According to former Netflix CFO Barry McCarthy: "They just about laughed us out of their office."

But that wasn't the end of the story. By 2007, Blockbuster were on the right track and had an internet movie component that was steamrolling Netflix. Netflix's management wanted to sell their company to Blockbuster to save face, but Blockbuster decided they didn't need them—their growth was strong.

In a strange twist, there was a boardroom dispute at Blockbuster later that year that produced a change in CEO. The new guy, Jim Keyes, had no clue where the future of the company lay. He decided Blockbuster should be a rental business rather than an online entertainment one. Bad idea. Keyes wasn't worried, though; in 2008 he would state:

"Netflix [isn't] on the radar screen in terms of competition."

Within eighteen months, Blockbuster had lost 85 percent its value. Within three years, they were filing for bankruptcy.

Blockbuster went belly-up and Netflix went on to thrive. With over 100 million subscriptions worldwide, Netflix changed the way many view their entertainment.

THE RISE OF NOKIA (2000)

The Nokia 1011 was released in 1992 and was the first commercial GSM (Global System for Mobile communication) phone. Earlier in the year, Jorma Ollila had taken over the company and pushed Nokia in the direction of telecommunications. He sold off other areas of Nokia to make this happen. Ollila had the right idea at the right time. Nokia had been in the red in 1991, but by 1999 they were making $4 billion in profit and had overtaken Motorola as the number-one mobile phone manufacturer.

In the year 2000, Nokia dropped the gauntlet. They released the indestructible 3310. It was the perfect phone for the time: easy to use, durable, good battery life, great button keys, and some addictive games to boot. The phone went on to sell 126 million units worldwide. It was iconic. My older sister had a bright red one; I loved playing games on it. The endless array of personalized covers was all the rage among the teenagers I looked up to.

The 3310 was just the start for Nokia. Other smash hits included the 6280 in 2005, and the 2007 Nokia N95, which had the first 5-megapixel camera to hit the United States, as well as GPS maps and more.

I remember the first phone I bought with my own money in 2004. It was the Nokia 6020, featuring a color screen, 3.5 MB of storage, and an amazing 0.3 megapixel rear camera (about 40x less resolution than the standard today). We may laugh today but being able to take videos on something so small back then was amazing to me.

First Camera Phone (2000)

Smartphone cameras are getting to the stage where they can replace introductory stand-alone cameras.

The thought for most today is: why lug a large camera around when I can use my phone? They're convenient and almost as good, so...

Parts of movies have even been shot on iPhones—that's the level we're at.

The first camera phone (phone with a built-in camera) was the J-phone by Sharp. It was released in Japan in November 2000. Its resolution was 0.11 megapixels.

In 2001, the BBC would report on the invention on website. Some of the comments are pretty interesting to look back at now:

"I love taking pictures, but I hate having to carry another gadget. If it is integrated with the phone, as should the organizer and the browser, that would be just so cool."

Gabriel Henao, USA

"A picture-shooting cell phone certainly is a curious invention. It could be handy for delicate investigation

or infiltration. If you disguise it a bit better, who would know to look for a camera on a phone?"

Johanna, Finland

"Take pictures of friendly dogs I see when I walk around."

John, USA

"It's an obvious move. Eventually all portable gadgets—phone, camera, palm computer—must come together in one communications device."

Julian Ilett, UK

Julian gets it; he's right on the money! Also, a little creepy there, Johanna.

IPOD MAKES APPLE A CULTURAL ICON (2001)

It was white, the size of a deck of cards, and it had a strange circular wheel and enough space for about 1,000 songs. No one knew what to make of it at first, but soon it would become omni-present.

Looking back at Steve Job's original presentation for the iPod, it is clear that Apple had the right idea. This is how they got there:

Digital cameras and camcorders were starting to enter the market, and digital music had seen an explosion in popularity, thanks to internet file-sharing services like Napster. To capitalize on this trend, Apple built a range of Mac "iApps" for photo, video, and music. During this process, they studied the available devices that would need to work with these apps, and realized that iApps understood the devices, but the devices didn't understand iApps.

Here was the question: What if there was a device made specifically to work with the Apple software? It would be a level of software and device integration that hadn't been seen before. Apple thought about starting with photos (capitalizing on the digital camera and camcorder trend), but in the end chose music—it was part of everyone's life, and Jobs loved music. There was no market leader yet, so it was prime time for Apple to get in.

The original iPod was a quantum leap. The iPod didn't just hold one CD like a Discman or flash (chip-based) mp3 player. Thanks to its thin hard drive, it held your entire music library in your pocket. The original iPod had a capacity of 5 GB. This was around half the capacity of a desktop computer in 2000, and it fit in your pocket! The iPod also had a more advanced battery than even the Apple laptops at the time, allowing for ten hours of battery life. But perhaps the biggest innovation was the scroll wheel.

At the time, mp3 players were difficult to use. You had to manually click each song. If you had a thousand songs…well, forget about it. How could you scroll through thousands of songs in a second? The scroll wheel changed all of that with its built-in acceleration, featuring an audible click and tactile feedback. It was a breakthrough in user interface design. What was once a chore had now become a joy. This was classic Apple.

The iPod was also tiny compared to everything else, considering the capacity it held. It was ultraportable, light, and bore that trademark Apple design. It wasn't an easy task, though.

I Don't Care, Just Make it Slimmer!

According to ex-Apple-employee Amit Chaudhary, here's a story about how Jobs got his engineers to slim down the original iPod:

When engineers working on the very first iPod completed the prototype, they presented their work to Steve Jobs for his approval.

Jobs played with the device, scrutinized it, weighed it in his hands, and promptly rejected it. It was too big.

The engineers explained that it was simply impossible to make it any smaller. Jobs was quiet for a moment. After a little while, he walked over to an aquarium, and dropped the iPod in the tank. When it touched the bottom, bubbles floated to the top.

"Those are air bubbles," he snapped. "That means there's space in there. Make it smaller."

After being barked at, the engineers achieved the impossible, and the original iPod was born. Initially, iPod sales started off lackluster, due to being Mac-only. However, sales began to take off when Tony Fadell, head of the iPod project, convinced Jobs to make the device work with Windows.

SPACEX (2002)

Elon Musk was born in 1971 in South Africa. After graduating from the University of Pennsylvania with degrees in business and physics, he co-founded the company PayPal. Shortly after its success, now a wealthy man, he quickly looked to the skies for his next endeavor. One day, he had the unusual idea to send a collection of plants to Mars, to inspire people by showing that humanity can do amazing things. When Musk set out to get quotes on the cost of the mission, US companies estimated a cost of around $80 million. Unimpressed, he tried to buy a refurbished Russian intercontinental ballistic missile, but the deal fell through.

Around the same time—in the early 2000s—NASA was pulling out from the business of launching spacecraft. Musk saw this as an opportunity, and in 2002, he founded Space Exploration Technologies Corporation or (SpaceX).

In the 2010s, SpaceX began to get a lot of media hype, but there is some basis behind the buzz. This was the origin; now, we'll take a look at what the company has achieved to date.

What Has SpaceX Achieved That Other Companies Haven't?

Milestones previously only achieved by government agencies:

- The first private company whose rocket reached orbit (2008)
- The first privately-developed rocket to put a commercial satellite in orbit (2009)
- The first private company to successfully launch, orbit, and recover a spacecraft (2010)
- The first private company to send a spacecraft to the International Space Station (2012)

Historic Firsts for Humanity

Historically, recoverable, reusable rockets are launched, fly into orbit, and return to earth gently (landing upright), have been the holy grail of rocket science. Reusable rockets significantly cut costs and lessen turnaround time. Such a feat was thought to be infeasible until SpaceX made it routine. Using this method, they accomplished:

- The first ever launch and upright landing of an orbital rocket back on earth (2015)
- The first relaunch and landing of a used orbital rocket (2017)

The company also plans to complete the first private-capsule manned mission to the International Space Station in 2019.

SOCIAL MEDIA RISES (2002)

Humans are social beings; most of us want to know more about each other.

There's a famous experiment that goes something like this: If you put one person in the middle of the street and make him stare up at the sky, everyone will walk past and ignore him—he's just a crazy person, right? But, if you put a group of people in the middle of the street and make them stare up at the sky, more people will stop to look—there *must* be something interesting up there. That many people just *can't* be wrong.

Groups of humans tend to flock to what everyone else is doing (a term I lovingly call "sheeping"). This intrinsic part of human nature was key for the success of what we now call "social media."

The first popular social networking site was TheGlobe.com. It launched in 1995 and went public in 1998. TheGlobe.com still holds the record for the largest first-day gain of any IPO. Its CEO Stephan Paternot became the poster boy for the excesses of dot-com millionaires.

But the pace of change in social media was rapid in the 2000s, and as TheGlobe and others would learn, it was difficult to find the winning formula.

Friendster (2002)

Friendster was founded in 2002 by Canadian programmer Jonathan Abrams. Before founding Friendster, Abrams was a Senior Software Engineer at Netscape, but left the company in 1998.

Around 2001, he started thinking of a new way for people to connect and interact, a way that would integrate their offline and online lives:

"The way people interacted online was either anonymous, or through aliases or handles," said Abrams. "I wanted to bring that real-life context that you had offline, online—so instead of Cyberdude307, I would be Jonathan."

He was onto something. Anonymity affects how people behave. At least initially, people tended to be more cautious on a site like Facebook, where there were real-life consequences to what you said. Contrast this with somewhere like YouTube today, where abusive comments are flung around like monkeys having fun with excrement. Sometimes it's best to just view the video and not the comments.

Abrams would solidify this idea using Match.com as a basis. He wondered, what if people's profiles gave information about them and linked to others, instead of focusing on dating preferences used for hooking up?

The idea was to model the types of connections that happen in the real world—a sort of friendship network. This idea would become Friendster, a play on the popular mp3 downloading site Napster. In 2002 friendster.com was registered and had three million users in a few months. It was clear that there was something powerful happening.

Google was impressed by what Friendster was doing. They offered $30 million to buy out Friendster in 2003, but the offer was turned down. At the time, Friendster was receiving a lot of attention in the press and it was the top social network...until April 2004, when it was overtaken by Myspace. Even after it began declining in the US, the site was doing well in Asia…until Facebook. It would all be over by 2009, and in 2015, Friendster closed its doors for good.

Myspace (2004)

Even if you didn't know about Friendster, chances are you knew about Myspace. During the middle of my high school years, Myspace was *the* big thing. Customizing my profile, browsing music, and choosing profile songs was a large part of my spare time. The "Top Friends" feature caused a lot of drama, because you could only choose eight. Perhaps that was a design flaw.

Myspace began in August of 2003 with a group of employees of the online marketing site, eUniverse. They all had Friendster accounts and liked the idea of a social network, so they decided to make their own. eUniverse threw its full support behind their project.

By 2004, Myspace had a large user base, mainly among teenagers and young adults. Its focus on music meant underground bands (particularly of the rock and emo genre) posted their work on their own Myspace pages. Bands like Taking Back Sunday would enjoy huge online popularity, thanks to this feature. In early 2005, Myspace began talks to acquire Facebook, but the offer fell through when Mark Zuckerberg asked for $75 million. The 100 millionth Myspace user signed up in 2006, and it was the leading social networking site in late 2007.

It wouldn't last, however. Myspace was overtaken by Facebook in mid-2008. Myspace's decline continued as the site stayed true to its music roots, refusing to adapt or add new features. Additionally, when all the teenagers grew up and went to university, Facebook—which was less tacky and better suited for adults—was waiting for them.

Facebook (2004)

The difference between Facebook and the previous social networks is that Mark Zuckerberg continued adjusting the social media site to keep people on it—for better or worse.

Born in 1984, Mark Zuckerberg was a bright kid, and this was clear even at an early age. Growing up as a millennial, he had experience with computers and learned programming from his father. Zuckerberg had already written his own software by middle school, and his dad hired a software developer to tutor the young programmer. Zuckerberg's tutor famously said that Zuckerberg was a prodigy, and it was tough for him to stay ahead of his student.

In high school, Zuckerberg and a friend built a media player called Synapse. It learned listeners' habits and recommended music it thought they would like. Synapse was such an intriguing concept that it gained the attention of Microsoft and other global players, but Zuckerberg wasn't settling; he wanted a bigger challenge.

By the time Zuckerberg graduated from high school and entered Harvard, he was a seasoned programmer. At Harvard, he wrote a program called Facemash, essentially a version of "hot-or-not" with a voting and ranking system attached. To the dismay of some less-attractive students, Facemash went live on the weekend and by Monday, the Harvard servers had crashed. The students couldn't even get on the internet—it was that popular.

Around this time, the second generation of social media was starting to take off, as Myspace and Friendster gained ground. Three Harvard students, Cameron and Tyler Winklevoss and Divya Narendra, had been working on a site called "Harvard Connection" for a year, and had recently lost two programmers to graduation. They were looking for a new, talented programmer to take over some of the work.

After hearing about the Facemash phenomenon, it was clear that Zuckerberg was their guy. Mark accepted their offer and hopped on board the Harvard Connection project, though he had his own ideas. Although the details are shrouded in controversy, it is said that Zuckerberg betrayed the Harvard crew by working on his own site, while continuing to delay work on the Harvard Connection. In the end, Zuckerberg took inspiration from the social idea of Harvard Connection, and the resulting effort was "The Facebook." Its name was later changed to just "Facebook" and the rest, as we know it, is history.

We can no longer think of Facebook as just a social network: Facebook Incorporated has grown much larger than that.

Today, Facebook has 1.4 billion active users; it is the third most visited website after Google and YouTube. Every 20 minutes on Facebook, 1 million links are shared, 2 million friends are requested, and 3 million messages are sent. Every day, Facebook creates 4 new petabytes of data with 350 million photos uploaded per day.

Facebook is worth over half a trillion dollars—that's more than Coca-Cola, PepsiCo, and McDonalds combined.

As Facebook grew, however, it became the thing to hate. Data-sharing scandals and negative effects on society have rocked the company. A number of studies have shown that it makes people miserable, and CT scans have shown that the social network visibly decreases empathy in developing brains. There'll be more on the social impact of Facebook in the final chapter.

YOUTUBE CONNECTS THE WORLD THROUGH VIDEO (2005)

A Little YouTube Tribute

Without YouTube, there would be no *ColdFusion*. There wouldn't even be this book that you're reading. I got my start on YouTube in 2009 by posting videos about the HTC HD2 phone running 3D UI Android software mods. The channel really took off when I showcased the capabilities of the original Samsung Galaxy Note. It was at a time when phones were still small, and a large-screen phone (then referred to as a phone-tablet) was very unusual. I immediately saw that this was the future: a phone you could freely browse files on, as well as browse the web and view desktop sites comfortably.

It was obvious to me that the more capable our phones became, the greater the need for more screen real estate. The problem with the original Note for me, though, was that it felt very much like a phone. I showcased some software hacks on YouTube that made it feel more like a mini-computer, which I thought was much cooler back then.

Others watching evidently thought the same. I was often laughed at by non-tech friends for carrying such a large phone and using it differently. Today, most smartphones are large. In my mind, the phone was evolving into something new, and I just showed people what was possible. I'm grateful that I could share thoughts and ideas with the world and people responded. So thanks, YouTube!

YouTube is an Underappreciated Privilege

When I was a kid, whenever I wanted to watch a video, it had to be on VCR (and later, on DVD). And then there was YouTube. Today we can watch a virtually endless array of content at any time, for free.

The YouTube phenomenon has definitely changed the world. Its sheer size and influence is mind-boggling. There are seven billion hours of video watched every month. It would take fifteen straight days of watching just to view the amount of footage uploaded every *sixty seconds*, and sixty-five years to watch all the footage uploaded each day.

YouTube is the second most-visited site on the internet, and the second largest search engine on the internet. That gives you a glimpse of how big YouTube is. But how did this all happen? Let me tell you a story.

YouTube Origins

It all started with PayPal, the company co-founded by Elon Musk. Three employees of PayPal would change the world: their names were Chad Hurley, Steve Chen, and Jawed Karim. After their days at PayPal, they were interested in creating a video-dating website. The result was called "TuneIn Hookup." It was a site where people could upload videos of themselves in order to find dates, and as you might guess, it didn't work out that well.

People, and especially women, didn't want to upload dating videos. The site was so desperate it began posting ads on Craigslist. Despite offering to pay women the extravagant sum of $20 to upload videos of themselves, nobody came forward.

But, as so often happens with these things, serendipity struck. One day in 2005, Karim noticed how hard it was to find Janet Jackson's 2004 Super Bowl wardrobe-malfunction video online. This sparked an idea. There had to be an easier way to upload and share videos. The technology existed—it was just a matter of building the right website.

Karim realized that they could use the existing code from their TuneIn Hookup site as the basis for a video website. TuneIn Hookup was modified to allow any kind of video to be uploaded.

Soon after, the website's name would be changed to YouTube, and on February 14, 2005, in a humble room above a pizza shop in California, the YouTube company was founded. YouTube's first video, 'Me at the Zoo,' was uploaded on April 23, 2005. It was a video of Jared Karim awkwardly talking about the elephant behind him.

It wasn't long before YouTube had its first million-hit video. This happened in September 2005. It was a Nike ad featuring Brazilian player Ronaldinho doing tricks with a soccer ball.

In October of 2006, Google stuck their heads in the door and thought, "Hey…that's pretty good…" They would acquire YouTube for a cool $1.65 billion.

Exceptional Growth

Google called YouTube the next step in the evolution of the internet. It was a pretty bold statement, especially as the YouTube company only had sixty-five employees. But Google knew YouTube was something big, something very different from what had come before. It quickly became one of the fastest-growing sites on the Web.

It was all in the name: The "you" in YouTube was in reference to the producer, and the aim was to broadcast *yourself*. This theme was highlighted in 2006, when *Time* magazine's *Person of the Year* was "You." The caption on the cover read "You. Yes you. You control the Information Age. Welcome to your world." This was true—we had the power to turn YouTube into anything we wanted.

In May 2007, the YouTube Partner Program came into the picture. This changed the lives of many people. For the first time, YouTube made it possible for everyday people to turn their hobby into a business. About a year later, the most successful users were earning six-figure incomes. It wasn't long before YouTube became truly mainstream.

During the exponential growth of the 2000s, the website produced its own living, breathing culture. It began to replace TV, which was amazing considering the short amount of time that it had been around.

YouTube is already bigger than TV among people under eighteen, according to the BBC. Some very real celebrities had their start on YouTube. One that comes to mind is Justin Bieber. There have also been shows, like *Broad City* and *Workaholics*, that got their start as YouTube skits before becoming successful primetime TV shows.

Today, YouTube is the single biggest distributor of video media on earth, from independent small films, to news media, to entirely new genres of programming. YouTube isn't just cats, vloggers, and viral videos anymore.

In 2010, Chris Anderson, the owner of TED Talks, noted that crowd-accelerated learning from YouTube could result in the biggest learning cycle in human history. Perhaps that hasn't happened yet, but I do know that I've learned a great number of skills, such as video editing, from watching tutorials on YouTube, and I'm sure you've have had similar experiences. It is common for YouTube videos to be shown in schools and universities around the globe. For the first time, mankind has a shared digital video archive.

However, some say the true essence of YouTube has been diluted recently, with corporate interests, advertising scandals, and demonetization running rampant. The website is in a state of flux at the time of writing, and it seems like things may change in the future. It will be also interesting to watch competitors like Twitch, and crypto-decentralized video sites.

THE SMARTPHONE ERA BEGINS (2007)

The smartphones of today are like mini PCs in our hands. The rate of growth of smartphones, in terms of power efficiency and processing speed, has been significantly greater than that of PCs.

In my 2013 *ColdFusion* video "The Greatest Story Ever Told," I stated:

"The 2010s will be the decade where the future meets today. We'll begin to see a few things. Firstly, smartphones will become incredibly powerful, at a rate which many find hard to keep up with. We'll see 64-bit mobile operating systems emerge. We'll see glimpses of the very distant future, one example of which is the convergence between the smartphone and the PC. Picture this: in five years' time, your smartphone being as powerful as your PC is today, running PC software."

It may be slightly over-dramatic but, at the time of writing this book, this is more or less true. The convergence has begun. In 2018, Apple announced that it's going to begin bridging the gap between iOS (mobile) and MacOS (desktop)—that is, allowing iOS apps to work on the Mac. By 2018, full Windows PCs that ran on the same chips as smartphones were being shipped. The rate of improvement I was already seeing in 2013, coupled with the sheer amount of funding that was being poured into faster and more efficient processors, had to result in such gains in a small form factor—keeping up with Moore's Law in the process.

The smartphone revolution is certainly interesting, but perhaps more interesting is the story of how it all began. Of course, that was with a shiny new iPhone.

iPhone (2007)

With over a billion units sold, the iPhone is the most influential product of all time. It turned the smartphone into an instant gateway to the world: everything you desire, in the palm of your hand.

It's All About Touch

Human-computer interaction was a mess before smart devices, and manipulating digital objects was a chore. In the early 2000s, zooming in on an image usually meant clicking on a menu, selecting the "zoom" option, and choosing the amount you wanted to enlarge it by. How we zoom in on an image today couldn't be more different. We now have the option of simply pinching the screen and manipulating it with our fingers.

Touchscreens existed in the early 2000s, but were predominantly resistive touch (that is, screens you have to press down hard on, like those in ATMs or the information screens at train stations and airports). On smaller screens, it's inexact and frustrating. In the '90s, Apple tried to use this resistive touch in the Newton, and it failed.

This was the first problem that had to be solved.

LG tried an all-resistive touch screen phone in 2006, but an ATM-style experience on a phone was as clunky as you would imagine. By 2007, the world was looking for a new approach to the mobile phone. They just needed a company to be in the right place at the right time, with the right product.

The first seed for the iPhone came, not from Apple, but a small company called FingerWorks in 1999. Founded by Wayne Westerman, FingerWorks had figured out how to use a different type of technology effectively—capacitive multi-touch. It was fast, responsive, precise, and most importantly, smart enough to recognize multiple fingers and what they were trying to do.

In the early 2000s, FingerWorks released a trackpad called iGesture. It helped people with wrist injuries use a computer without aggravating it, as Westerman suffered from a wrist condition himself. He even wrote a simple AI program to help the system understand the differences between accidental and intentional touches. When gestures were performed, the pad would interpret them and turn the movements into computer shortcuts like copy, paste, and scroll.

FingerWorks would play a critical role in the development of the iPhone.

MP3 Cell Phones Threaten Apple

In 2004, the iPod was making up 50 percent of Apple's sales, but there was a cause for worry. The company became increasingly concerned that mobile phones would destroy iPod sales. Phones could now play mp3s and were becoming a lot like iPods. Something had to be done.

Early on, Steve Jobs hated the idea of Apple making a phone. He was concerned with a lack of focus in the company (this was the precise issue that Jobs had solved upon his return to Apple) and believed that a phone would only serve a niche geek market. Cell phones at the time were not the easiest to use, so I can understand why it wouldn't have seemed right for a hip company like Apple to get into this market.

Around the same time, Tony Fadell, who led the successful iPod division, suggested to Jobs that it would be a good idea to put WiFi in an iPod. While Jobs thought about it, Fadell and his small team got to work on a new iPod—a PDA hybrid prototype. The result was...a disaster. Imagine an iPod with modified software allowing users to navigate the web...with a click wheel. Jobs hated it. He understood that it *worked,* but thought it was a garbage experience. Jobs told Fadell to try another way.

A Touchscreen Mac?

At Apple, a group of engineers and software designers would meet weekly in what had been a user testing room. They were from different departments, but united by curiosity and imagination. They realized that the web and digital revolution was bringing richer and ever-more-complex media to computers and that clicking and typing might not be the best way to navigate that future. What if there were a more fluid way to interact with content? With this idea in mind, they created an informal media-integration group within Apple.

One day in 2002, an Apple employee by the name of Tina Huang brought a FingerWorks device to work, because she'd hurt her wrist. It was a black rectangular pad that allowed the seamless execution of complicated computer tasks, by just the use of the fingers. This way of working took the strain off Huang's wrist. More importantly, for this story at least, the pad had smart multi-touch built in.

The FingerWorks pad was seen by some of the Apple employees from the weekly meetings in the user-testing room. Already thinking of new ways of media-interaction, to them, multi-touch interaction was an interesting prospect.

Inspired, they whipped up a demo to show the Apple marketing department. Using the multi-touch gesture pad and a projector, they displayed an interactive image of MacOS X. The idea was that you could use finger-based gestures on the pad to manipulate elements of the desktop MacOS software. The demo was met with minimal enthusiasm by the marketing team; they didn't see a use for it. How could they? There was no product that would really *need* it, except perhaps an advanced in-house Mac Mouse.

Unfazed, the media interaction group continued have weekly meetings to work on the possibilities.

Johnny Ive, one of the weekly members from the informal media-interaction team, had also shown Jobs the concept. Initially, Jobs rejected it. He remarked that it would only be good for "reading something on a toilet."

After further thought, Jobs warmed up to the idea, and the project was greenlit. The multi-touch MacOS project did trundle along, but it became riddled with problems, and was ultimately shut down.

In July 2004, Jobs had surgery to remove a malignant tumor in his pancreas. The realization that he might have limited time on this planet helped accelerate the timetable of what needed to be done at Apple.

We Need to Do a Phone

In 2005, Bas Ording, a programmer and UI designer, received a call from Jobs asking him to make a demo of a touch interface accessing a scrolling contact list. Jobs now wanted to do a touchscreen phone.

While Ording was making the demo, he noticed that, when scrolling through a list, the text image would suddenly stop when it hit the top of the screen. Because of this sudden stopping of all motion, he thought his code had crashed. After some time, Ording noticed he had in fact reached the end of the list. It gave him an idea: Why not have the image "bounce," so there's some visual feedback that lets you feel like you've reached the end of the list—not like it's suddenly crashed. This was the genesis of the "rubber-banding" effect on the iPhone. Ording was a Gen-Xer and would later state that playing games like *Super Mario Brothers,* with its simple animations that just "felt right," gave him the inspiration.

When Jobs saw Ording's rubber-banding effect for the first time, he realized that a phone *could* indeed have a touch interface.

While work on the scrolling list demo was happening, other touchscreen projects were secretly going on at Apple as well. As cool as these demos were, they were no more than just a bunch of disjointed concepts: some pinching and zooming, a few widgets like stocks, notes, and calculator…but none with a unified structure.

Developing a Product: 2005

Jobs wasn't impressed with the disjointed array of demos—it didn't seem like there was a product to sell. In early 2005, Jobs gave the team two weeks to create something great, and it had to be great… or else.

A small team at Apple spent two sleepless weeks trying to get the company's first touchscreen phone right. They focused on the vision of a phone and its function: How do you make a phone call on a touch screen? How would you get from a calendar to Web browsing? What's the logical flow of getting from one application to another? Amazingly, by the end of the two-week period, they had something to show Jobs.

It's Show Time

The first time Jobs saw the prototype, he didn't jump for joy, or exclaim anything. He was silent. Sitting back, he said: "Show it to me again." He was, in fact, blown away. The project would be top secret within Apple from that point onward. Around the same time, Apple purchased FingerWorks, bringing FingerWorks' whole team on board to help figure this jackpot out. At one point, forty people within Apple were working on touch technology alone.

The Battle Within Apple

Apple had two choices on how to transform this technology into a product: enlarge the already successful iPod into a phone or shrink MacOS down into a phone using FingerWork's touch technology. Nobody knew which would work best, so Steve let both ideas run.

The iPod-enlarging team was led by Tony Fadell, and the multi-touch MacOS team was led by Scott Forstall. Forstall's team were seen as the underdogs. After all, Tony Fadell had helped push millions of iPod sales and was also working on two smash hits—the iPod Nano and iPod video.

The battle had begun, but the thing was, neither team was allowed to know what the other was doing. In fact, the hardware guys weren't allowed to see any software, and the software guys weren't allowed to see the hardware. Not one person at Apple knew what this phone was going to look like until the keynote…and they weren't given a date for that, either.

Apple's iTunes Phone Catastrophe

While these projects were going, top Apple executives convinced Jobs that something had to be done about encroaching mp3-capable mobile phone sales. Jobs agreed to partner with Motorola (the hippest phone company at the time, with their thin Razr phone) to make an iTunes mobile phone.

Essentially, it was a way for Apple to not cannibalize iPod sales (their previous fear), while giving people a chance to try iTunes and hopefully buy an iPod. Apple would have no involvement in the hardware, only focusing on iTunes integration.

The result was the Motorola Rokr. Already outdated on release, the thing just...sucked. The Rokr was such a dumpster fire that it was soon being returned at a rate six times higher than the industry average. Consumers were expecting something big from Apple, and this wasn't it.

To the shock of Motorola executives, Jobs decided to subsequently unveil the iPod Nano in the same presentation. The Nano was an impossibly small, color iPod that served to sweeten the bitter taste of the Rokr. When I first saw it revealed on that stage in 2005, I thought it was the most beautiful consumer electronics device ever made.

After the Rokr failure, Jobs returned his attention to the iPod phone (Tony Fadell's team), as it was the safer option, while still leaving the Scott Forstall touch-MacOS team to continue.

An iPod that Makes Phone Calls

Fadell's team tried a plethora of designs, one of which was similar to a video iPod, but with a phone mode. When you wanted music, it would behave like a regular iPod, with touch controls for pause, play, etc., around the scroll wheel. When you needed to dial a number, you would switch it into phone mode and the device would behave like a rotary phone. Jobs still insisted that he could make this concept work.

This design had the ability to browse WiFi and also had a built-in speaker. As with earlier efforts, it was clear that a scroll wheel was an obstacle. When using the device for more complex functions like dialing a number or texting, the scroll wheel would just make things more complicated. Interestingly, Jobs originally thought a "back" button was necessary for function, but the design team insisted that simplicity in function would make the consumer trust the device more.

Which Software?

Nobody within Apple really knew what this new device actually was. There were arguments between the two teams. The iPod team saw it as another portable accessory like an iPod, so software wasn't important. The touch MacOS team saw it as a full-fledged, multi-touch computer that fit in your hand.

Fadell and his team were certain that this new phone should run a beefed-up version of the iPod software, while Forstall's team thought a shrunken-down version of MacOS X would be better. Forstall's team theorized that mobile chip technology had become powerful enough to run a version of MacOS X. Soon they managed to get scrolling to work smoothly on a compact version of MacOS, and from this point, it was decided that this would be the way to go. This shrunken MacOS X would become known as iOS. Soon, Fadell's iPod-phone idea would be abandoned, and all efforts would be focused on iOS.

As the software began to be written, the details had to be ironed out. How do you unlock the thing without doing it by accident in your pocket? One day, Freddy Anzures, an Apple user-interface engineer, found the solution in the most unlikely of places: a toilet...

One day Anzures was on a US domestic flight and felt the need to relieve himself. He got up from his seat and went to the toilet. As Freddy locked the cubicle door, he happened to observe the locking mechanism. It was so simple, you just slid it to unlock. This concept lit a light bulb. The result was the famous original slide-to-unlock feature.

Meanwhile, Jonny Ive was beginning to imagine what the hardware of this phone could look like. Although later designs would stray, the first sketch from Jonny Ive was close to the final product. Ive was imaging an infinity pool, "this pond where the display would sort of magically appear. The rest of the device had to get out the way."

Interestingly, Ive didn't want a headphone jack in the original phone.

Difficulties Mount

As the project dragged on, tempers rose within the company, due to the secrecy and competition and the time pressure of the project. And it was no wonder. Everything the teams were making was new. Touchscreen technology was in its infancy. The Apple engineers had to figure out how to make a transparent version of multi-touch mass-producible. Untested custom chips had to be developed. Cell reception had to be worked out. Material designs needed to be perfected.

Basically, these teams, which had never made a phone before, were now trying to make the most ambitious device ever imagined. Talk about running before walking! Imagine how hard it must have been for the developers, too. If an app they were working on crashed, it could be because of virtually a million things, from their coding to any one of the fresh hardware components.

By early 2006, iOS was making great progress in the phone's development, but the keyboard still sucked. Scott Forstall paused all development of applications for iOS and made everyone focus on the keyboard issue. Everyone on the team built keyboards for three weeks, until one employee finally came back with an accurate keyboard. The breakthrough was made by using primitive AI to figure out what was being typed (essentially "predictive text"). For example, if you were to type the letter "t," there was a high chance that you were going to type "h" next. So, the keyboard would make the contact region for "h" larger, without changing the appearance of the button to the naked eye. Although it will probably remain a mystery who at Apple came up with the keyboard breakthrough, my hunch is that it was Wayne Westerman of FingerWorks. He had already mastered the multitouch/AI synergy, after all.

Down to the Wire

As it was all coming down to the wire for Apple, disaster struck. Trying to get all the best names in for their breakthrough phone, Apple had partnered with Samsung to create the phone's custom CPUs, but Apple didn't give them a lot of time. Just three months before the launch, the custom CPU chips from Samsung still had bugs that caused the phones to crash. With so little time left on the clock, it was looking like a catastrophe in the making.

Other aspects of the iPhone were still being slapped on *very* late in the development cycle. One of its killer apps, Google Maps, was only added as an afterthought. On the hardware side, the original iPhone screen was supposed to be plastic like the iPod's, but the decision to use glass was made one month *after* the launch!

The Day that Changed the World

When the day of the keynote came on January 9, 2007, none of the teams within Apple knew exactly what the final iPhone product looked like, even though they had worked on the project for years.

When Jobs took to the stage, some things were still incomplete. The iPhone's buggy CPU issue hadn't been properly solved—only patched up for the demo. This meant the phone could crash any time during the presentation.

The Apple team held their breath, sweating as they sat and watched Steve Jobs excitedly proclaim "...And we're calling it...iPhone!" There was a mostly positive crowd reaction, but also some uncertain laughter, as some thought Jobs was joking about the name. Yet, the demo was going well, and the crowd was loving the smooth scrolling and the technological magic they were witnessing. As Tony Fadell, leader of the iPod-phone team looked on, Jobs did something truly cruel. When showcasing to the crowd how to delete a contact, he gleefully slid Fadell's name off the iPhone's contact list. Jobs was basically saying Tony was fired. Forstell states that during the rehearsals of the presentation, Jobs would always delete a random contact, never Fadell's. To the crowd, it was just a demonstration of how this phone made everything fun and cool to use. To those working at Apple, it was a message. Fadell was in trouble.

This one act was a summary of how the project had been for most—brutal. The project was so hard on the secret teams that it had ruined marriages and cost some workers their health.

Oblivious to the sacrifice, the world was electric with the buzz of the iPhone when all was said and done. Keen fans lined up to get one.

The week after the iPhone was announced, I remember chatting with a friend who was nine years older than me. I told him about the fact that there was a phone coming out with just a screen and no buttons, which you navigated strictly by touching. After seeing Jobs' demo of scrolling and gestures, it made sense to me that this was the perfect interface for a small-screen device. I distinctly remember him throwing his hands up in the air and in a mocking tone saying "OOoooOOooo, wow! No buttons!" A couple of years later, he had an iPhone.

The future had arrived; finally, a computer in your pocket. As Apple software engineer Henri Lamiraux put it: "We took a Mac and we squished it into a little box. It's Mac Two, same DNA, same continuity. NeXT to Mac to iPhone."

It had thousands of times more computing power than NASA did in 1969 when they put man on the moon., yet was super easy to use, with a full web-browsing experience to boot. In a nod to Engelbart's original graphical-interface vision, even children could pick up an iPhone and know what to do. Capacitive multitouch and gestures had broken down walls in human-computer interaction.

Other small touches—like a proximity sensor that turned off the screen during a phone call, so it wouldn't hang up if it touched your face by accident, or an accelerometer that could sense if you were holding the phone in portrait or landscape mode—made the whole package *feel* like magic.

What followed the iPhone was a complete shift in how we thought about technology. Now information and news were instant. We became totally connected all the time. In the next decade, this over-connectedness would cause some issues.

The iPhone

There's an App for That: The App Boom (2008)

Once the initial fanfare was over, the sales of the iPhone slowed. It would take the introduction of the App Store in 2008 for things to really kick off again for Apple.

If not for the breakthrough of the App Store, the iPhone could have just been one of those devices that looked cool and was cool to use, but never caught on. Initially, Apple marketed the concept of web apps. The plan was for the user to open up the Safari Web browser and go to a website that was designed for iPhone. From there they could find HTML-based games, utilities, and applications. Obviously this didn't go over well, because Apple would soon announce the App Store as a response. Before this, for an entire year after the iPhone's launch, it could run a grand total of sixteen apps. There was only one home screen of apps and that was it—not to mention how limited the iPhone was compared to some other devices out there. It had no multimedia messaging, no video camera, no cut and paste… the list goes on.

Nokia, Blackberry, and the rest of the established phone companies had been providing a lot of these features for years, but the difference was that for the first time, a device was now a changeable blank canvas with no buttons. Since the device could have apps run on it, the iPhone software creators could make it be whatever they wanted it to be. Their creativity was only limited by their imagination and the hardware available. Today, billions of apps have been sold, and the mobile app space has become a new industry.

Companies like Nokia, Blackberry, and Palm thought they knew the game, but they couldn't visualize the future. They thought things would never change, and before they realized, it was too late to prosper in the brave new world. A new way of thinking was needed to compete, and there was one company up to the task—Google.

Enter Android (2008)

In 2003, midway through the iPod storm, something new began to take form. Andy Rubin and Rich Mena founded a company in Palo Alto, California, named Android Incorporated.

Android was initially supposed to be a digital-camera software company. Early on, it was clear that this market was way too small, so the company concentrated on producing a smartphone OS that could take on Nokia's Symbian, and Microsoft's Windows Mobile. They believed it was too difficult to get new products out to consumers in a timely fashion, and concluded the missing link was an open platform. And so, they created a smartphone OS that had one.

In July 2005, just after purchasing YouTube, Google poked their head in the door again and thought: "…Hey, that's pretty good…"

Google would acquire Android for over $50 million. With this capital, Android began developing prototype phones. The early Android prototypes were more like Blackberries—they had no touchscreen and a physical QWERTY keyboard. Later prototypes were reengineered to support a touchscreen, competing with other devices such as the 2006 LG Prada, and of course, the iPhone, in 2007.

Android played a major role in the late 2000s.

The next few years became a blur as the world we know today began to form. Just like the battle between the Macintosh and the Windows PC in the mid- to late-'80s, the game was on once again, though this time the tussle was between the iPhone's iOS and Google's Android operating systems. The first Android phone arrived in 2008. It was the G1. The G1 was the joint effort of Google and HTC. It wasn't bad for a first try, but as expected, there was a lot of room for improvement. The iPhone ran circles around the G1 in terms of speed, ease of use, and code efficiency.

But Android had something up its sleeve. Android's open-source and fully customizable software was key, as it meant that any individual or company could use Android as a basis and tailor it to their liking. This proved to be very effective, and soon, Android software would be running on everything from phones to cars to fridges to portable movie projectors.

As the 2010s drew to a close, the world was transformed once more. We were connected by social media like never before; all of human information was in our palms and at our fingertips. In the next decade, we saw science and technology shift into hyper-drive.

Impossible Possibilities
The 2010s

In the times of the Greek philosophers or scholars of old, it was reasonable for someone to strive to know everything that was ever known. The web and the internet have made this beyond impossible. By 2017, 90 percent of all information had been created in the previous two years. This eerily echoes Vannever Bush's vision in chapter 7 of a knowledge and information explosion. Knowledge and its distribution among some of the most brilliant minds on the planet began to bring forth a rapid advancement—not only in science and technology, but the world at large. This progress began to show strongly all over the world in the 2010s. In this final chapter we'll see how. But first, a bit of background.

In the 2010s, the developing world began to catch up at breakneck speed. Between 2000 and 2012, the poverty rate of the world fell by half—the fastest period of economic development in all of history. Rates of child mortality in Africa became lower than those of Europe in 1950. Some of the fastest-growing economies in the world were in sub-Saharan Africa, where millions began to have access to predominantly Chinese-made smartphones. With the addition of satellite internet, the world was at their fingertips, too. India experienced a digital revolution in 2017 with cheap and fast mobile data from the Reliance company. Renewable energies picked up steam in the decade due to a renewable-energy price drop and a global concern for climate change by world leaders. In the same vein, solar-powered aircraft were proven viable with the introduction of Solar Impulse 2, which circumnavigated the earth in 2016. Globally, battery tech began to improve, allowing the drone industry to explode overnight.

Innovation became a more global affair, with small start-ups popping up all over the developed world. Even many of my friends have gone on to be involved in start-ups. From the start-up scene, novel forms of personal transport began to emerge, including hoverboards or self-balancing boards, electric bicycles, and electric skateboards. Affectionately called "the Mouth Fedora" by artist Post Malone, electronic cigarettes or "vaping" became a cultural marker. The sharing and gigging economy came online with early entries like Uber and Airbnb. Smart glasses and smartwatches began their life cycle as a product category.

In the 2010s, fashion embraces the hipster subculture. Beards and moustaches gain popularity, and the undercut hairstyle reigns supreme. Superhero films and sequels do well. CGI animation remains strong, and *Toy Story 3* in 2010 becomes the first animated film to make more than $1 billion worldwide. House music, trap, and EDM see a massive return, and pop music at this time sees many elements borrowed. Internet meme culture has also affected music, with PSY's "Gangnam Style" and Baauer's "Harlem Shake" seeing billboard success. The maturing of online distribution outlets leads to the fracturing of music as a whole. Chillwave and bedroom producers are able to reach a new audience. Platforms where anyone can now upload and share mean electronic music has begun to split into ever smaller genres, known as microgenres, including Glo-fi, dream-beat, bedroom pop, and dream pop. All you need is a laptop and the internet.

Against this backdrop, the 2010s sees a parabolic acceleration in technology. All the previous decades had come together to serve as a launchpad for this decade. To see a relatable example of this rapid pace of tech, you need look no further than your pocket. A smartphone released six months ago is hopelessly outdated. A consumer product cycle this fast has never been seen in history.

This pace of advancement didn't come without a price. Phone addiction became a real issue among the millennial generation and Generation Z. The psychological effects of social media and the internet on the minds of young people, especially children, became an increasing concern during this decade.

To give some perspective, here's a quote from Henri Lamiraux, the man who oversaw the software engineering of the iPhone, the most influential device of our time: "I see people carrying their phone everywhere all the time. I'm like, okay, it's kind of amazing. But, you know, software is not like my wife, who does oil painting. When she does something, it's there forever. Technology—in twenty years, who's going to care about an iPhone?"

In 2018, both Apple's iOS and Google's Android announced software features to encourage digital well-being—features to help use your phone less. There'll be more on the price of advancement later.

Ever-stronger gains have been seen in mobile technology and laptops due to a dramatic increase in power efficiency. My 2017 Dell XPS 15 is a thin and light laptop with 32 GB of RAM—unthinkable just three years before its release. Toward the end of the decade, laptops (PCs) with 128 GB of RAM and phones with 16 GB of RAM are possible. Contrast this with the year 2000, when average PC RAM was 1 GB.

A storage revolution has also been on the rise. Hard-drive tech at the start of the decade resulted in huge, bulky power bricks with only 500 GB capacity. Eight years later, 8-terabyte hard drives are thin and light. Even phones had 1 TB of storage by this stage.

Cyber hacking and data leaking would increase in the decade. Edward Snowden shocked the world when he proved that the NSA did in fact have global surveillance. The WannaCry ransomware in 2017 took cyber-hacking to a new level: 230,000 computers in 150 countries were hacked. The only way to get access back was to pay the hackers in Bitcoin. All sorts of entities, from hospitals to airlines to FedEx, were targeted.

Despite privacy concerns, smart-home systems such as the Apple Homepod and Amazon Alexa have become mainstream in the decade. Robotic pets make their debut with the Anki Vector pet. E-sports explode in popularity, even being considered for the 2020 Tokyo Olympics. In the latter part of the decade, the tech companies become king. Facebook diversifies by acquiring Instagram for $1 billion, and WhatsApp for $19 billion. Both become massive social media platforms. The continued growth of e-commerce companies such as Amazon forces the closing of traditional shopping malls. Google is fined a record $5 billion EU for violating antitrust laws. Apple becomes far and away the most valuable publicly-traded company on earth.

China begins to rise to prominence in this decade. They land on the moon and overtake the United States as the world's largest trading nation. They even file more patents than the US. In GDP ranking, China went from ranking below one hundred to number two in thirty-five years! One of the reasons for China's growth was its new status as the production hub of the world. The output was not high in value, but it was ubiquitous; everywhere you turned there was a "made in China" label. Outstripping even China's economic growth was India, the fastest growing major economy in the world currently. Meanwhile, the once powerful Japan saw a credit-rating downgrade (the nation is seen as less likely to pay back debt). In addition, over 20 percent of Japan's population is over the age of sixty-five, making it the most elderly population on earth. To deal with this, robots were quickly developed to handle the demand in elder care. The rest of the world saw a massive number of retirees from the baby boomer population causing financial strain through pension programs. In fact, in the United States, 10,000 baby boomers hit sixty-five per day!

The global economy was in a sluggish recovery since the global financial crisis of the late 2000s. Some economic analysts argued that the problem was not fixed at all, only patched over with low interest rates and central-bank money-printing. The latter only inflated the stock and housing markets instead of creating jobs, causing a slow and silent squeeze on the middle class of western nations. The economic pressure would also lead to civil conflict in some nations. The 2015 Syrian civil war caused a refugee influx in Europe, the largest since World War II. Germany took in the largest number of immigrants, creating social tension and political instability in the nation.

Economic woes also affect the United States. In 2016, the US presidential election split America, as a large number, dissatisfied with the economic situation, voted for Donald Trump. The echoes of surprise, anger, and confusion rippled through the globe, and the left- and right-wing ideologies became polarized. The ability of social media to amplify the voice of the individual only intensified the divide. Extremes of either political side are the loudest, often reinforcing the opposing side's stereotypes.

Correct information became a valuable resource as the post-truth and "fake news" period took hold. Vannever Bush could not have foreseen this side effect of his vision. Mainstream media has weakened and trust is at an all-time low, with the media employing clickbait tactics to cut through the information noise. This has further fed polarization and resulted in the overall loss of confidence in governments and institutions.

At the start of the decade, everyone was entitled to an opinion. But by the end of the decade, everyone is entitled to their own truth, reality, and facts. Nuance dies; reality gets blended. Those desperate to make sense of the suddenly fracturing social fabric are driven to modern philosophers like Jordan Peterson, Sam Harris, and Bret Weinstein. This group was named by Weinstein the "intellectual dark web."

Other notable events include the discovery of the Higgs Boson (the thing that gives matter weight) in 2012, the Curiosity Rover landing on Mars (2012), the ability to manipulate specific memories (in mice) using optogenetics (2013), the detection of gravity waves, the experimental use of graphene in batteries, and the James Webb telescope, which can see exoplanets (planets outside our solar system). As a side point, the first exoplanet was discovered in 1988—we didn't know if they existed before this. By the mid-2010s, scientists had discovered thousands of them.

The Mobile Grows (2010)

In 2010, Apple, Inc. launched its first tablet computer, the iPad, which offers multi-touch interaction. It was the last new product that Steve Jobs oversaw before his death from pancreatic cancer. The iPad became an immediate hit and, only months after its release became the best-selling non-phone tech product in history. Before the iPad, tablet computers were a mess, just as touch-screen phones were before the iPhone. Android tablets followed shortly after, and soon PC sales were declining in favor of tablets and convertibles. Netbooks would become extinct as a consequence.

In the US, people were spending more time on mobile apps than the World Wide Web by 2011. Globally, 4.6 billion people had a mobile phone subscription. The next year, Samsung overtook Nokia to be the largest phone maker. The Chinese phone manufacturer Huawei overtook Apple as number two in 2018. Also in 2012, tablet and smartphone sales overtook the Netbook, which died off later in the decade.

The 2014 Nintendo Switch first demonstrated, on a massive scale, the newfound power of mobile chips. The Switch was a breakthrough product which could be used either as a handheld gaming system or plugged into a TV for a full-fledged console experience. The device was Nintendo's fastest-selling gaming system, and the fastest-selling console of all time in the US and Japan, selling over 14 million units within a year of its release.

3D Printing Matures (2010)

3D printing is advancing rapidly, providing new materials in every field from engineering to medicine. As the technology matured, it was utilized in everything from printing houses to human organs, to food and patient-specific prosthetics. Here's a question though: If 3D printing has been around since the 1980s, why are all of these things just popping up in the 2010s? It has to do with the expiration of a key patent in 2009. As soon as this patent expired, it was off to the races for innovative manufacturers keen to explore this new world of 3D printing.

To give you an idea of the rate of progress, let's compare today's 3D printing technology with that of the time of my thesis. In 2012, my thesis topic was 3D printing (selective laser sintering) with stainless steel material. The printers at my university were state-of-the-art, but the size of four refrigerators, and cost almost $1 million. By 2018, steel 3D printers could fit on a desktop and cost around $120,000.

3D printers for plastics and polymers are even smaller—around the size of a microwave—and much cheaper, at under $400.

In South Korea, a new technique for 3D-printing human skin has been developed by scientists. Their new method is around fifty times cheaper than alternative methods and requires ten times less base material. Great news for those suffering from burns, or in need of skin grafts.

In 2018, researchers at an American university successfully 3D-printed glass. The "liquid glass" they designed is a glass powder embedded into a liquid polymer. Researchers at ETH Zurich 3D printed the first ever entirely soft artificial heart in the same year.

The Rise of Machine Learning through AI (2012)

With the exponential increase in computational power, machines gained the ability to recognize objects and translate speech in real time using a technology called machine learning (ML). ML powers the voice recognition in our phones and smart-home systems, the image recognition in self-driving cars, and Facebook's ability to detect your friend's face. In other words, machines are getting "smart." ML in the 2010s revolves around the concept of teaching computers to learn like humans. The ML impact is far wider than most people think.

Despite how it seems to the average person, no major technology exists today without ML being built into the heart of it. ML controls internet traffic, electrical grids, code compiling, fruit picking, baking, banking, radiology, phone and server farm power management, antenna design, oil exploration, terrorist detection, police assignment, border policing, issuing traffic tickets, and so on. Most economically important tasks where computers used to be in charge were replaced by ML in the 2010s, and it's been exponentially progressing from there. In the mid-2010s, a field of machine learning began to make massive strides in doing what we once thought impossible. We'll take some time here to make sense of ML and related technologies, understand their applications, and investigate the impact they are having on our world.

PART 1: MAKING SENSE OF IT ALL

Difference Between Artificial Intelligence, Machine Learning, and Deep Learning:

For most people, the terms AI (artificial intelligence), ML (machine learning), and DL (deep learning) are interchangeable. Before continuing, it's important that we make some concrete definitions.

Artificial Intelligence

Artificial Intelligence, or AI, is the umbrella term for a branch of computer science. Its aim is for machines to mimic human cognition with a focus on complex problem-solving. True general AI should be able to do whatever a human can (and beyond). This is the ultimate goal of this technology.

Machine Learning

Machine learning is a subset of AI. ML focuses on how to make machines learn on their own without the need for hand-coded instructions. ML systems analyze vast amounts of data, apply their knowledge, train, and learn from previous mistakes, then complete a specific task effectively. Some interesting things began to happen through the middle part of the decade, however.

Google's "Alpha Go" ML system showed the ability to do tasks outside what was intended. More on that later.

Deep Learning

Deep learning is a subset of ML. This technology attempts to mimic the activity of neurons in our brain (particularly the neocortex, the wrinkly 80 percent, where thinking occurs). This setup is known as an Artificial Neural Network (ANN). ANN systems do learn, in a very real sense, although the way they learn is not yet fully understood.

The basic idea that computer software can simulate the brain's large array of neurons in an artificial "neural network" is decades old, but it wasn't practically possible until the 2010s. We'll get to that story in a bit.

The Potential of Deep Learning

In 2012, Microsoft chief research officer Rick Rashid wowed attendees at a lecture in China with a speech software demonstration using deep learning. The software transcribed his spoken words into English text, with an error rate of only 7 percent, subsequently translated them into Chinese-language text, and then simulated his own voice uttering the text in Mandarin. In a few years, this feat would be commonplace.

It goes further than this: Imagine a program that could guess what's happening in a given scene just by listening to the audio. Not only this, but it's consistently better than humans at guessing. What about predicting human behavior (greeting, waving, hugging, handshakes, etc.) only from watching sitcoms all day?

Or a program that speaks in a voice so realistic that it is indistinguishable from a human?

This is the power of ANNs, and by the latter half of the decade, these technologies were very real. In a few years, these aforementioned feats will also be commonplace.

The rate of progress in this may take you by surprise, but you would be in good company. Even Google's co-founder Sergey Brin was in awe of the entire phenomenon of deep learning. Here's a quote from Sergey in 2017:

"I didn't pay attention to it at all to be perfectly honest…
[H]aving been trained as a computer scientist in the '90s

everybody knew [deep nets] didn't work. This revolution has been very profound and definitely surprised me, even though I was right in there. It's an incredible time and we don't really know the limits. Everything we can imagine and more. It's a hard thing to think through and has really incredible possibilities."

How Do Artificial Neural Networks Work?

To really understand Artificial Neural Networks, we must first look at ourselves.

When we are exposed to certain stimuli (data such as an image, a sound, or touch), chemical signals are sent to our brain and certain neurons fire. The firing of these neurons depends on a minimum threshold. This basically means that if the signal reaching the neuron is strong enough, it will fire (be activated), and if the signal is too weak, it won't. It's basically like an "on" and "off" switch. Yes, just like transistors and the binary language of computers. Furthermore, our brain can change and adjust the connections between our neurons by strengthening existing connections, or removing unused connections, or even creating new connections. This is the power we have been given. When we learn something, the paths between neurons are strengthened.

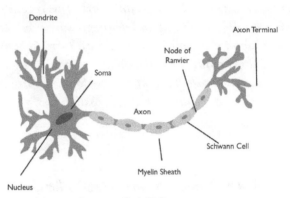

Structure of a typical neuron

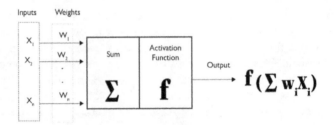

Structure of an artificial neuron

- An artificial neuron takes this basic idea and makes a simplified caricature of the process.

- The nucleus is replaced with a node (which is a mathematical function that tells it when to activate).

- The dendrites are replaced with inputs (incoming data).

- The synapse is replaced with a weighting function (which determines the likelihood of an artificial neuron firing, a number between 0 and 1).

- The axon is replaced with an output (data that represents an answer).

Our brain's 86 billion neurons are organized in a complex 3D structure, allowing for an almost limitless array of connections. An artificial neural network, on the other hand, is arranged in layers, with an input and output layer and "hidden layers" in between. The hidden layers are where the learning happens.

Let's solidify this concept with the task of recognizing a face: A network may break the image down into elements within each layer. For example, the first layer detects shading, the second layer, outlines, the third, shapes, and so on. For the curious, the mathematical method of calculation simply boils down to a set of massive matrix multiplications, where the goal is to minimize something called a loss function. A loss function is essentially a measure of how wrong the system is. The weights that determine when neurons fire are adjusted based on the loss function. Minimize the loss, and your neural net (ANN) will be pretty good at its task.

Artificial Learning

To appreciate how these systems learn, we must look closely at how we're able to tell the difference between similar things at all. We need to stand outside our perspective of being a human for a second.

Let's take cats and dogs as an example: When you think about it objectively, they are both four-legged furry animals with ears on top of their heads… So why don't we ever confuse the two? As children, we're told which animal is which by someone who knows. When we understand that a certain animal is one thing or the other, we strengthen the paths between the neurons that fire when performing that task. Very quickly, we stop needing new examples. Our learning process is robust enough to correctly identify a new cat that looks different from one we've seen before. Artificial Neural Networks do a similar thing for an image-recognition task. They're given specific examples of cats and dogs, certain neurons fire, and then it's told if it's right or wrong (by its loss function score). As a result, certain connections strengthen (weight values are increased), and they learn with each example. This process is called "training."

Now that we have an understanding of what Artificial Neural Networks are, let's see the story of how this turned into the field of "deep learning."

History of Deep Learning

It all started with psychologist Frank Rosenblatt in 1957, who developed what he called a "perceptron." A perceptron was a digital neural network that was designed to mimic a few brain neurons. Rosenblatt's first task for this network was to classify images into two categories. He scanned in images of men and women as data, and hypothesized that, over time, the network would learn the differences between the two (or at least see the patterns that made men look like men and women like women). Again, this is essentially how we learn.

Just a year later, the hype for Rosenblatt's technology was strong. In 1958, the *New York Times* reported the perceptron to be "the embryo of an electronic computer that will be able to walk, talk, see, write, reproduce itself, and be conscious of its existence."

Unfortunately, despite the hype, Rosenblatt's neural network system didn't work very well at all. He used only a single layer of neurons in his perceptron, making it extremely limited in what it could do. There wasn't much that could be done about this, as computers at the time could only handle one layer. In 1969, Rosenblatt's colleague and childhood friend Marvin Minsky published a book that assessed these experiments. In it, Minsky stated that neural networks were "generally disappointing" and "hard to justify." The book was said to have so impacted the computer science community that the idea of Artificial Neural Networks was abandoned for over a decade.

The work may have been discarded, but a keen computer scientist by the name of Geoffrey Hinton thought everyone was plain wrong. He theorized that since the human brain was indeed a neural network, it evidently made for an incredibly powerful system. Artificial Neural Networks *had* to work somehow—maybe they just needed some tweaking. To Hinton, writing off the idea seemed like a colossal waste.

Originally from Britain, Hinton set to work on his idea at the University of Toronto in 1985. He and his team quickly realized that the problem was with Rosenblatt's single-hidden-layer approach. What was needed were more layers in the network to allow for greater capabilities. Serendipitously, computers were now powerful enough to handle it. Today, we call this multilayered approach a "deep" neural network. With some tinkering, the artificial neural net began to work. In the following years, ANNs did two things never before seen.

Dean Pomerleau built a self-driving car for public roads in 1989. In the 1990s, a neural net system that could successfully recognize handwritten numbers was built. As exciting as this was, there wasn't much more that could be done with the technology. It just didn't perform well in much else. Another dead end was reached.

This time the problem was two-fold:

- A lack of computing power.

- Insufficient available *training* data.

- ANNs were called wishful thinking once again. Hinton, who was once again outspoken, didn't give up and continued pushing for further research. He would be ridiculed for his long-standing faith in a failed idea and sidelined by the computer science community.

Fast-forward to 2012, and computers began to be fast enough to make Artificial Neural Networks practical for wider applications. In addition, the data from the internet meant that these systems could now be robustly trained. The stage was set for ANNs—all that was needed was for someone to prove their worth.

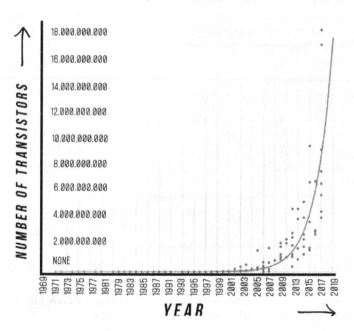

The Birth of the AI Movement

The real birth of the modern AI movement can be traced back to one event in 2012: ANNs bursting on the scene by proving themselves in the ImageNet challenge.

Founded in 2009, ImageNet is essentially a database of real-world images that are used to train machine-learning systems. The images are in 1000 categories ranging from cars, animals, and buildings to abstract concepts like love and happiness. ANNs were never before tried in this challenge.

A large data set was the final missing piece in the puzzle to make ANNs finally shine. This increase in data for training caused ANNs to begin working. Consider again the problem of recognizing a cat. If you only saw three cats, you'd only have a few camera angles and lighting conditions to work with. Something as simple as seeing the cat from a different angle, or with different lighting, would throw you off. But if you'd seen a thousand cats, you'd have a much easier time recognizing one. This is the importance of data. Computers aren't as intuitive as humans and need more examples. In the field of AI, data is the essence that allows machines to learn. If data is a new form of oil, ImageNet was a huge oil reserve, and the only machines that could run on it with astronomical success were ANNs.

By 2012, the ridiculed Geoffrey Hinton was now sixty-four years of age and for him, continuing the work wasn't an easy task. Hinton had to stand all the time due to a back injury that caused a disc to slip if he ever sat down. He and his team created the first Artificial Deep Neural Network to be used in the ImageNet challenge in 2012. Hinton's program was called AlexNet, and it had performance like no one had ever seen. AlexNet destroyed the competition, scoring a success rate of over 75—41 percent better than the previous attempt. This one event showed to the world that Artificial Neural Networks were indeed something special.

After this point, everyone began to use ANNs in the ImageNet challenge, and the accuracy of identifying objects rose from Hinton's 75 to 97 percent in just seven years. For context, 97 percent accuracy is surpassing humans' ability to recognize objects!

Computers recognizing objects better than humans had never happened in history. Soon the floodgates of research and general interest in ANNs would change the world.

Image recognition is now commonplace, even recognizing disease in medical imaging. As for ImageNet, it caused training data sets to blossom. Training sets grew for video, speech, and even games.

PART 2: MACHINE LEARNING APPLICATIONS

So, by this stage, you're probably beginning to see how powerful machine learning is. But just how far can they go? In 2012, a deep learning AI named IBM Watson won the game of Jeopardy by answering to human natural speech in real time. The AI did so by reading vast amounts of information on the internet and found patterns to make sense of it all.

In 2016, Google's AlphaGo created a watershed moment by beating world champion Lee Sedol at the world's hardest game: Go. Although its rules are simple, this ancient Chinese game has enormous complexity. It isn't in the same realm as chess, where a computer can just calculate all the moves. In fact, there are more possible moves in Go than atoms in the universe, and it's said to require human intuition to play. When questioned why they made a certain move, professional players would just state that it "felt right." For this reason, Go has been the holy grail of AI. AlphaGo wasn't taught the rules of the game; it learned by watching other people play. The algorithm then played against versions of itself until it had reached a superhuman skill level. Later on, AlphaGo showed the ability to perform tasks outside what it was originally intended to do.

By 2018, IBM had an AI that could debate humans, and Watson was working with doctors to give medical advice. It did this after reading and making sense of millions of medical studies.

AI in Medicine

With an increasing number of baby boomers living longer globally, doctors may soon need some help. A Chinese deep learning system called Miying served 400 patients just one year after its release in 2017. Artificial intelligence is a key part of the government's "Made in China 2025" plan, which aims to make the nation a global leader in high-tech industries, including robotics, by then. Some in the computer science field are envisioning a revolution in specialized healthcare. For example, the wealth of data from millions of wearables could train an AI to be useful in aiding your personal doctor.

An interesting study was done on human ophthalmologists, and it found that humans are statistically poor at diagnosing certain eye diseases. In experiments, any two human ophthalmologists will only agree with on a diagnosis 60 percent of the time. Worse still, if you give any single ophthalmologist the exact same image that they read a few hours earlier, they'll only agree with their own diagnosis 65 percent of the time. By 2018, there were Artificial Neural Networks that outperformed human ophthalmologists. By this stage, some ANNs were capable of early detection of esophageal cancer with an accuracy rate of 90 percent. That was with less than a year of operation. Google is also experimenting with deep learning in retinal images to provide early detection of diabetic eye issues. In addition, deep learning ANNs were found to be 10,000 times faster than radiologists.

Other Applications

In 2017, Deepfakes made the process of replacing the faces of a person in live footage just a click. Previously, this was a tedious process costing hundreds of thousands of dollars. All you need now is a collection of a facial images from Google, and the deep learning ANN does the rest. Below is a movie scene in which Nicholas Cage's face imposed on a female actor's body.

In 2018, the Nvidia company created an algorithm that made extremely choppy, slow-motion video smooth, as if shot with a specialized super-slow-motion camera. In the same year, the company unveiled an ANN graphics card that could calculate ray-tracing in real time. Ray-tracing is used in computer graphics to calculate how light is reflected and absorbed by certain materials. Ray-tracing is the reason why CGI movies (think *Toy Story*) look so different from video games. Nvidia's ANN took most of the hard work out of the calculations by heavy optimization. Thanks to ANN's, desktop users and gamers could experience cinema-grade computer animation—something once reserved for Hollywood studios.

ANN's can also dream up a sequence from a still image. Imagine this: You give an ANN a still photograph, then it creates a segment of video footage of what it thinks should happen next in the scene. If there's a photo of a beach, the waves tend to ripple next; if there are people in a scene, they start to walk.

Further ANN capabilities include:

- Converting black-and-white images and video into color automatically.
- Photorealistic image generation from just a typed description
- Given a video sample, turning night into day and vice versa, or changing the weather.
- Enhancing a pixelated image (making it clear), just like in CSI .
- Error correction in space telescopes, and the list goes on.
- This leads us to the big question. If these systems are progressing in capabilities exponentially, what happens to us? This is commonly called the automation problem.

PART 3: A DISASTER OR THE NEXT INDUSTRIAL REVOLUTION?

Effects of Automation: The Robots Took Our Jobs!

With all the progression in technology, there has been fear of disruption to jobs. In the 1920s, there were guards who opened and closed the doors of subway trains and elevators. These have been automatic for more than a generation. When was the last time you saw a doorman? Since the 1950s, 8 million farmers, 7 million factory workers, and hundreds of thousands of telephone operators have been replaced by automation. In the 1960s, President Kennedy ranked automation as the number-one risk to employment. Yet, work still persisted.

This Time It's Different

When you really think about it, there's no clear limit to what can be automated with AI. And that's the crucial distinction between the current and past technological revolutions. At minimum, almost every task that isn't creative, or doesn't require a ton of mechanical precision or super-detailed human interactions, will be automated in the foreseeable future. This will have profound impacts on the way the world functions that are obviously very difficult to predict.

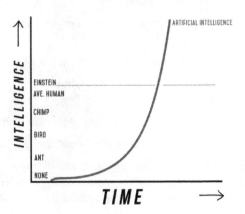

I think such things will cause social discomfort as they begin to encroach upon jobs. There are both positives and negatives to the rise of ML, but what do we do with all this continual shift in possible impossibilities?

During the next decade, it is thought that many mundane tasks could be handled by AI, creating more free time for people to do creative things. I've heard this argument before. I would imagine that not everyone on this earth is creative or would have the time and money to pursue this.

Another thought is that advanced AI could increase productivity, creating new industries and new suppliers to support those industries. All of this would mean new jobs. Views on this differ, with some experts highlighting the need to research financial solutions to AI disruption. This includes concepts such as universal basic income, based on a trustless system like blockchain. (More on blockchain later in the chapter).

Perhaps, in the future, instead of working for a company, it will be more common for people to run their own businesses with their own personal AI acting as a productivity aid. But I don't believe this is a widely held view either. Globally, an estimated 800 million jobs will be at risk of automation by 2030.

When AI is talked about, some people automatically think of a Doomsday Skynet scenario. But personally, I think what we should be more interested in is who controls the AI. If the strong AI of the future is open-source and available to everyone (with safeguards of course), it's possible that this could be a good thing. Conversely, if the best AI on earth is held by a few, the probability of a good outcome drops significantly. It would be a case of whoever owns the AI makes the rules, in the latter scenario. To stop AI getting into the wrong hands, institutions such as the Future of Life Institute and Open AI have been set up. It shows that people are already thinking about the next best steps for the future.

Virtual Reality and Augmented Reality (2012)

Imagine a future where you could put on a headset and be in a totally different world, interacting with objects, talking to people, and living a completely second life. Or playing a video game where you *are* the character, not just moving it around. This is the promise of virtual reality. Virtual, from mid-fifteenth century English, means "being something in effect, though not in actuality or in fact." The basic idea is to simulate or replace our reality with a computer-generated or otherwise different one.

When you put on a headset, for all intents and purposes you're somewhere else, from the point of view of your brain. You're no longer watching something, you're inside it. Consider that, compared to print, film has the extra elements of vision and sound. In the same way, VR has the extra elements of physical motion, space, and presence. For this reason, it has been called the first empathetic technology. For the first time, it allows someone to *feel* what it's like to be another person. This was highlighted in the 2014 experiment by the BeAnotherLab project. It allowed a male and female to experience what it was like to be the opposite sex, using VR.

VR had been tried for the better part of a century, and most notably in the 1990s, without much success. The computing power and display technology just wasn't there. This all changed in 2010 when Palmer Luckey, a seventeen-year-old kid, built a prototype virtual-reality headset in his parent's garage. He was frustrated with the low quality of existing headsets and decided to build something much better. The breakthrough was the use of the mobile-phone display technology in VR for the first time. This allowed for a clearer, wider field of view. The garage prototype eventually became the Oculus Rift headset. The innovation from Luckey kicked off a wave of next-generation VR headgear, though it would take most of the rest of the decade to be commercially viable.

Augmented Reality

"Augment" is from the Latin word *augere*, which means "to increase" or "to add." Augmented reality (AR) is like VR but involves a person's real environment being supplemented or augmented with computer-generated images, usually motion-tracked.

AR broke onto the scene with Pokemon Go in 2016. The mobile app was a poor demonstration of AR, but it became the technology's first widespread use. By the end of the decade, AR toolkits made virtual 3D objects as simple as scanning in an object with a phone. Adobe's Project Aero had software to make AR creation simple. Apple would push AR technology late in the decade. The company's interest was an indication that AR was coming to the masses.

CRISPR Gene Editing (2012)

CRISPR is a revolution that has seized the scientific community. Research labs worldwide have adopted a new technology that facilitates making specific changes in the DNA of humans, other animals, and plants, within a few years of its debut.

What Exactly Is CRISPR?

CRISPR (for Clustered Regularly Interspaced Short Palindromic Repeats) is a biological DNA system found in bacteria. Essentially, bacteria use the CRISPR system as a defense. CRISPR cuts up the DNA of invading viruses that could otherwise destroy them. It's like a built-in pair of biological scissors.

Better still, the CRISPR scissors can remember a virus, search for its particular DNA sequence, find a match, and then cut it. It's a tiny, automatic system programmed into biology. Pretty cool. The question now was whether scientists could get hold of this.

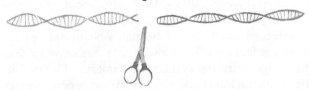

CRISPR technology was first discovered in 1987 by Japanese scientists, when they accidentally cloned a fragment of CRISPR along with the rest of their experimental material. They studied the system for a while but didn't observe the cutting and seeking-out properties. The full function wasn't understood until 2012.

What Can CRISPR Be Used For?

After discovery, CRISPR was soon being used to cut up, edit, and paste an organism's DNA code. Compared to previous techniques for modifying DNA, this new approach was much faster and easier.

Researchers have modified stem cells that may then be injected back into patients to repopulate damaged organs using CRISPR. One study injected modified stem cells directly into the brains of stroke victims. Most victims could regain major movement, and one even began to walk after being wheelchair-bound for many years.

Another study reengineered cancer cells to seek out and destroy blood-borne cancer or tumors. Swarms of the CRISPR molecular scissors would snip up and corrupt the cancer's DNA until it couldn't multiply.

In another cancer-fighting approach, CRISPR was used to make the immune system itself fight the disease. In many cases, the body's immune system doesn't recognize cancer, so CRISPR research took aim at this issue. Scientists reengineer the immune system to seek out and destroy cancer DNA.

In 2018, a study published in *Science Translational Medicine* demonstrated cells containing advanced CRISPR bacteria that triggered their own self-destruction after performing a certain task. These CRISPR bacteria cells managed to kill off a range of cancer cell types, while leaving healthy cells in the vicinity unharmed.

This biological method is relatively new, so the risks and long-term effects are not fully understood. However, this technique could offer faster, cheaper, and more precise cell therapies in cancer treatment.

Robotics Takes a Leap Forward (2016)

In 2016, the world was taken aback by robotics firm Boston Dynamic's YouTube video. The video showed a humanoid robot (Atlas) walking, picking up boxes, placing the boxes on shelves, and effortlessly correcting itself when pushed by a human. In 2018, more videos of Atlas surfaced. This time it was jogging, doing backflips, and performing many other tasks thought to be impossible just a few years ago. It seemed that suddenly robots had the athleticism of humans.

The breakthrough in robotics took so long because, as it turns out, moving is hard! Our brains have three to four times more neurons for mobility than cognition. We have around seven hundred muscles in our bodies. It just takes a lot of processing power to balance and be agile. With the help of faster computers and smarter software, a new type of robot had abruptly arrived.

Another robotics breakthrough occurred in 2018, but this time it was from an unexpected source: Disney. In their theme park, the Disney company had built what they were calling their "Avatar robot." It displayed the most fluid robotic motion the world had ever seen. In a promo video, the robot talked to the camera in a close-up shot. There were many YouTube comments asserting that it was CGI. It blew straight past Uncanny Valley for the first time.

QUANTUM COMPUTING BEGINS (2016)

Quantum computers have been called the next leap in human civilization—as big as toolmaking, farming, and of course, the Industrial Revolution. They are considered possibly as important as, or more important than, AI.

The essence of quantum computing is hard to grasp at first, but once understood, you'll see how this technology has the potential to change everything. You'll also understand just how bizarre reality is. For this reason, we'll spend some time here and explore the strange world of quantum computers.

Why the Fuss?

Imagine for a moment that there was a new type of computer that could calculate every possible solution to a problem simultaneously. You could give this computer a problem with almost infinite possibilities, and it could come up with an answer in a matter of hours. For perspective, if every computer ever made worked together on this same problem, it would take them many billions of years to calculate.

We call these contemporary computers "classical computers," and this new, seemingly impossible computer a "quantum computer."

How Does It Work?

While regular computers process data as binary bits (represented as either a 0 or a 1), a quantum computer processes data as quantum bits ("qubits"), which can exist in both 0 and 1 states at once, allowing multiple computations to happen simultaneously. This simultaneous state of both 0s and 1s is kind of like a soup of probabilities (called superposition), and it's a quirk of quantum physics (also known as quantum mechanics). This makes it possible to store and manipulate a vast amount of information using a relatively small number of particles.

To put it another way, a regular computer tries to solve a problem the same way you might try to escape a maze—by trying every possible corridor and turning back at dead ends, until you eventually find the way out. But superposition allows the quantum computer to try all the paths at once—in essence, finding the shortcut.

Backtracking a little, let's explore exactly what a qubit is. While a normal computer's bits are made from tiny transistors, a qubit can be anything that exhibits quantum behavior: an electron, an atom, a molecule, or even a loop of metal (if the environment is right— more on this later). If we take electrons as an example, they tend to exist in one of two states: spin up or spin down. If we call spin up a 1, and spin down a 0, in binary, we can begin to write code with it. Again, with superposition, it can be both a 1 and a 0 at the same time. Some researchers theorize that superposition is proof of alternative universes.

If you think this is all very strange, you're not the only one. The father of quantum computing, Richard Feynman, is quoted as saying: "If you think you understand quantum mechanics, you don't understand quantum mechanics." We'll get to much stranger quantum behavior later. But first, let's see how Feynman came up with this wild idea.

The Father of Quantum Computing

Born in New York City in 1918, Richard Feynman was definitely a unique guy. He was a theoretical physicist who worked on the Manhattan Project (under Vannevar Bush) in WWII, had a show on the BBC, was an excellent lecturer, and was one of the pioneers of quantum electrodynamics (a field of quantum theory that describes how matter and light interact). Besides developing the concept of nanotechnology, he was also one of the pioneering minds in quantum computing.

Feynman taught himself mathematics early on and received a perfect score on his entry exam to Princeton University. In addition to his formal education, Feynman taught himself concepts far ahead of what he was learning in class.

Feynman had other eccentric quirks. During his time on the Manhattan Project, out of boredom, Feynman cracked the codes of all the safes in the compound. One day he left a cryptic note in one of the safes that housed classified documents on the atomic bomb. When his fellow physicists found it, they were terrified. They thought German spies had stolen the secrets to the atomic bomb! Of course, it was just a Feynman practical joke.

In 1982, Feynman gave a lecture called *Simulating Physics with Computers*. In it, he stated a remarkable idea: Regular computers will never be able to model real-world quantum systems, no matter how powerful they get or how many are used together. Here's what he meant: Say you want to model a molecule to see how it behaves when it interacts with other molecules. In the real world, a molecule is formed when the electron orbitals of its component atoms overlap.

SHARED ELECTRONS

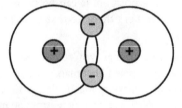

To accurately model real electrons, you need to keep track of the fact that they can exist in multiple states at *once*. Although this fact can be expressed as a probability (chance), it ends up being a real problem for classical systems (regular computers). This is because, as the number of particles goes up, the number of possible states (or probabilities) grows exponentially. Hence, for four electrons, there are 16 possible states. For 10 electrons, we would need to track 1,024 states. There's a total of 1,048,576 possible states for just 20 electrons.

If we want to a model a real physical system with *millions* of electrons, things get out of hand! In the classical computer world, a laptop can model a 26-electron system, while it takes a supercomputer to model a 43-electron system. And how about a 50-electron system? Forget it; that's impossible for any classical computer as far in the future as humans exist.

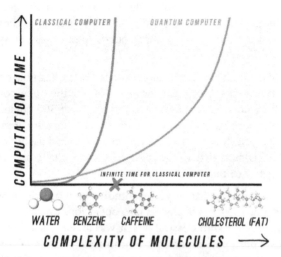

Nature (made from atoms with electron orbitals) and reality itself are quantum systems, and they can't be modeled on classical computers without making poor approximations. For the two simple molecules below, a classical supercomputer is off by a factor of 70 to 200 percent.

Molecule	Name	Bond Length		Error
		Experimental (Real)	Calculated	
CaF	Calcium monofluoride	1.967	4.079	2.112
Na2	Sodium diatomic	3.079	2.379	-0.7

It all boils down to this: The information required to describe a quantum system can only be *held* by another quantum system!

PART 1: CODING WITH NATURE

Enter Quantum Computers

Because of their qubits, quantum computers are quantum by design—just like nature. They have no problem keeping up with nature's exponential complexity. In other words, for each qubit added, a quantum computer gets exponentially more powerful, unlike a regular computer.

For the example below, the red graph represents a classical computer and the orange graph represents a quantum computer. Simulating the water molecule is possible for regular computers, but something as simple as caffeine is impossible—quickly leaving our abilities behind. To model caffeine and more complex models, even up to cholesterol, all we need to do is add a few more qubits to our quantum system—say, fifty or less.

With 100 qubits, it is theoretically predicted that a quantum computer would have more computing power than all the supercomputers on earth combined! A number register made of 300 qubits of data could hold more numbers simultaneously than there are atoms in the known universe. Imagine if future quantum chips had billions of qubits, a number comparable to classical computer CPUs and memory. Mind-boggling to think about.

Feynman suggested that, if you wanted to simulate the world, you could encode the rules of physics into operations on qubits, like using logic-gate circuits with classical bits. It was an incredible idea: almost like coding pure physics into the fundamental essence of nature and reality—not just some mathematical approximation of reality (like we do now). Next, we just hit play on the hyper-realistic simulation and see what happens. For example, we may discover a material that gets ten times stronger when some obscure molecule is added to its structure. Many theorize that we could invent new materials, new drugs, and novel energy-generation technology by modeling molecules in this way.

PART 2: QUANTUM WEIRDNESS

To get exponential speed-up, qubits must be able to communicate. They do this in the strangest of ways. In the world of quantum physics, the fate of all the qubits is linked together in a process called quantum entanglement. Entanglement is just a relationship between two super-positioned particles: the state of one determines the state of the other. Imagine it like this: You have two dice—one in Vancouver, Canada, and the other in Perth, Australia. If you rolled them 1000 times you'd expect them to roll completely different numbers. But if they're quantum-entangled dice, you can check the number 1000 times and every single time, it'd be the same. This weird phenomenon really bothered Einstein, who affectionately gave it the name "spooky action at a distance." He thought that maybe this was just something we haven't discovered yet. But this seems not to be the case.

In 2017, Chinese scientists demonstrated that quantum entanglement was occurring even if one particle of a connected pair was launched into space. It goes much further than this, however.

Once entangled, quantum particles stay that way, even if they are at opposite ends of the universe! In quantum computers, entanglement along with superposition helps the system store all possible solutions at once.

This may all sound bizarre, but when you think about it, the way we perceive the world in our everyday life is just plain wrong. Quantum mechanics is how nature really works. It's the foundation for the stability of matter, radiation, and magnetism. It *is* reality!

PART 3: MODERN QUANTUM COMPUTING

After Feynman, this whole quantum-computer thing remained just an academic curiosity, until 1994. American mathematician Peter Shor of Bell Labs found a theoretical way to use quantum computers to break codes based on the factorization of large numbers into primes. This doesn't sound like much at first, but many online security systems—from banking to encryption—rely on this. The underlying principle is that it's almost impossible, currently, to take a very large number and figure out what its prime factors are. The best algorithm known for classical computers uses a number of steps to crack the code, and this keeps increasing exponentially with the size of the number, hypothetically taking billions of years. Using Shor's algorithm, a quantum computer could perform the task in a few hours. From this point on, it was clear that quantum computers could do much more than model reality.

Not So Fast!: Quantum Design Problems

In the early 2010s, we began to build early quantum computers, but there were two main issues which remain today:

The first issue is due to quantum physics itself. The answers that come out of quantum computers are in the form of a probability. In order for a quantum system to "push" toward the right answer when asked repeatedly, the code is designed so that the qubits are more likely to be in the correct quantum state for the problem given (hence giving us the right answer). Fascinatingly, the quantum code is designed to use the wave-like properties found in particle physics to cancel out wrong answers and amplify correct answers. This answer can then be detected and viewed.

Think of a quantum computer's data like a light bulb randomly shining light around a room. The light particles are like qubits of information, and the quantum code is a set of (carefully positioned) microscopic mirrors. In this analogy, if we run a quantum computer, the light turns on, and the quantum code is written so that it causes the wave-particles of light to bounce off the code's tiny mirror surfaces, focusing the billions of rays of light to form a pin-sized spotlight on the floor. As this happens, the rest of the light in the room dims down. The spotlight represents the correct answer, and the dimmer light in the rest of the room contains the wrong answers, plus random noise in the system.

The second problem is the practicality of maintaining a quantum environment, which is why there aren't powerful quantum computers now. Qubits are like very shy divas. To be in their quantum superposition (multiple states at once), they need to be comfortable. Qubits must be shielded from all radiation and kept at a temperature just above absolute zero. If the qubit particle interacts with anything, the quantum effects are scared away. Any slight disturbance, such as light particles, radiation, or even quantum vibrations, can snap the particles out of their superposition state, voiding the entire advantage of the machine. As discussed earlier, a qubit is any form of matter that exhibits quantum behavior. This means that metal loops can be used to create the ideal environment. The older and more established qubit technology uses a metallic superconductor to create a superposition state. The

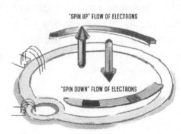

Super-cold metallic ring exhibits quantum properties

most popular method is a super-cold metallic ring (almost at absolute zero). At these temperatures, electrons can flow around without ever bumping into anything. Stranger still—all the electrons can flow clockwise, but at the same time, ALL the electrons can *also* flow counter-clockwise!

Today's efforts can keep a quantum superposition in effect for only a tiny fraction of a second—not long enough to carry out a useful algorithm. The maintenance of this fragile quantum state remains the major challenge for quantum engineers and scientists.

Quantum Computers and Real-World Problems

As it turns out, quantum computers are good at things that have a small **input** and **output**, while having a vast array of **possibilities**.

For example, finding the *longest* distance between two cities in Japan: The **input** is the map data (road distances, etc.); it's large but nothing a classical computer couldn't handle. The **output** is just one number, the longest distance. The **possibilities**...? It seems like a simple problem, but if you think about it, there's almost an infinite number of routes you could take.

In fact, this class of problem is forever unsolvable using current computers. There are many other real-world problems of this type—from weather patterns and climate change to boosting AI training, to modeling complex molecules, to looking at patterns in stock market data, or in recordings of brain activity. Unlike with today's computers, finding these patterns could be completely routine and require a less detailed understanding of the subject with quantum computers: an interesting idea, if used with caution.

A Quantum Future?

In the early 2010s, quantum computers were big and bulky, filling most of a room, with complicated cooling equipment creating the delicate environment needed for quantum effects. We've come full circle, as this is what mainframe computing was like back in the 1950s. If this is where we are now, could quantum computers decrease in size the same way regular computers have? Possibly, but not for a very long time.

Progress lagged in the industry until around 2016, when the pace of breakthroughs accelerated. IBM announced a 5-qubit machine and made it freely available for cloud computing. In the following two years, hundreds of research papers were published by scientists using the new technology. It was theorized that a minimum of fifty qubits were needed for quantum computers to do things faster than any classical computer can. This threshold is called "quantum supremacy." In 2018, Google released a quantum chip with 72 qubits (but most are there for error correction).

Although we are still a long way from practical quantum computers, the potential promise they hold cannot be overstated.

I think Feynman (who died in 1988) would be pleased with the renewed interest and progress we're starting to make in quantum computing. It takes a great mind to foresee such potential in the purely abstract.

Electric Aircraft Arrive (2017)

As battery technology improved, partly due to the new demand for electric cars, new possibilities became available. Batteries could power small, production-scale, fixed-wing planes.

A practical electric plane is something that has been elusive, even more so than practical electric cars. Battery technology has now improved to the point where a fully electric aircraft is feasible for some applications, such as flight training and short-distance travel. In 2018, I got the chance to spend time with one of the first commercially available, fully electric aircraft (the Pipestrel Alpha Electro), to see what the state of the art was like. The plane weighed only 350 kg, including its batteries, which can be swapped out in five minutes. The plane was so light, I could literally push or drag it around the hangar.

Unlike a regular petrol light aircraft, which uses fifty-year-old engine technology with hundreds of moving parts, the Alpha Electro had only one moving part in the engine. It also had only an on switch and a throttle lever to operate the engine. Older aircraft mix fuel on the go to operate the engine. The plane was quiet, and very easy to maneuver when I had a go at the controls. Electric aircraft have recently been considered for the air-taxi space. Established companies like Airbus and Boeing tried their hand at making vertical take-off and landing (VTOL) aircraft. These aircraft take off like a helicopter, but transition to horizontal flight for short distances. In 2017, the Chinese company E-hang began electric VTOL flights in Dubai.

Shortly after the 2008 global financial crash, a mysterious white paper emerged on the web. It was written by an unknown entity, a person or group calling themselves Satoshi Nakamoto.

In this mysterious paper, a new peer-to-peer financial system was discussed. This system was to use a digital cryptocurrency called Bitcoin and would be called blockchain. It was a boon to people sick of centralized power meddling with economic systems. Many perceived the system as a better way of doing things. Cryptocurrencies entered public perception in 2017, when Bitcoin rose in value by 1,000 percent and Ripple (another cryptocurrency) by 36,000 percent. Blockchain is being called the next stage of the internet by many.

Our current internet is one of copied and distributed information, be it video, email, or data. In other words, when you watch a video on YouTube or receive an email, you're looking at a copy of the original. When it comes to assets with real value, like money, votes, digital identities, stocks, bonds, or house or car ownership contracts, we don't want these things to have many copies. For example, you don't want to send someone a copy of $5,000 while you still have the original $5,000 under your name. This is called the "double-spending problem" by cryptographers. It basically means having two digital copies of something that should only ever have *one* unique identity.

The cryptography (mathematical security) of blockchain technology made it the first of its kind to solve the double-spending problem. Now, a new stage of the internet, based on the real value of real-value items, was available. This is partly why it's such a big deal. Some blockchain-based currencies promised anonymity from peering eyes. This gave rise to the Dark Web, an anonymous market for illegal drugs, firearms, and services, all powered by blockchain.

Simply put, blockchain is a continuously-updated list of who holds what. The list of records, known as a distributed ledger, is decentralized, available to everyone in the system to see and verify. This list is split into linked blocks that are secured using cryptography. The cryptography part means you can be sure about the records. This is dubbed automated trust, or a "trustless" system, meaning the trust is inherent, built into the system. Blockchain cryptography provides a mathematical way to confirm a participant's true identity. Due to its distributed nature, blockchains are unhackable with current technology. Everyone in the system can clearly see when a block has been messed with, and the compromised block can be automatically rejected.

While blockchain is the underlying technology behind the first wave of cryptocurrencies, like Bitcoin and Ethereum, its uses go beyond intellectual tech-savvy criminals, and even cryptos themselves. Blockchain has a feature called smart contracts. This allows for a real-world contract to be written in code, and to execute itself when certain conditions are met. This feature gave rise to blockchain-based social media websites and online games, but also saw use in banks and the NASDAQ. Smart contracts can be linked together to create entire automatic companies, largely replacing the need for a human management team in some cases. In addition to blockchain, other trustless systems, such as Tangle and Hashgraph, were developed later in the decade.

5G Networks Begin (2018)

We all have heard of 3G and 4G networks. 5G is a much faster network that promises to transform many aspects of modern life. Its main technical advantage is the use of much higher frequencies than current systems, allowing for more bandwidth. 5G will have 100 times more capacity than 4G networks. A two-hour film on 3G takes 26 hours to download; this is 6 minutes for 4G and on 5G, get this— just 3.6 seconds!

Response times are also much quicker. 4G responds in around 50 milliseconds, 5G, 1 millisecond. That's 400 times faster than you can blink. This means that a lot of work can now be done seamlessly in the cloud, something that wasn't quite possible before. Due to the low latency, self-driving cars would be safer, as every millisecond counts in an emergency situation. Analysts predict that 5G networks will become the skeleton for the internet of things: smart homes with connected appliances, integration into home security and lighting, self-driving cars, and industries that use medical equipment and industrial robots. Experts have claimed data rates of up to 100gb per second may be needed for these emerging technologies in the future. The aim is to have 1 million devices per square km supported. Say goodbye to slow internet at concerts, festivals and conventions. Gamers will rejoice as multiplayer lag will be a thing of the past. 5G signal towers can be focused to beam on areas with greater data needs, unlike

current systems, which radiate the signal out in a circle around the
cell tower.

Estimates predict 5G to reach 14 to 65 percent adoption by 2025. Adoption is currently held back by infrastructure development and cost. Unlike 4G networks that could use existing equipment, as it was on a similar frequency to 3G, 5G networks are of a much higher frequency (similar to satellite broadcasting) to accommodate all the extra bandwidth. This requires totally new infrastructure.

Currently China is taking the lead, with companies like Huawei and ZTE beginning 5G trials. 5G is planned for the 2020 Tokyo Olympics sites. Small trials have also begun in the United States.

5G Dangers?

5G has some people worried. The wavelength of the signal is approaching the tiny size of biological structures. A study from the University of Helsinki analyzed the reactions of human skin subjected to 5G frequencies. The study concluded that human skin will absorb the energy, especially the sweat ducts, but other organs remain unaffected. The health effects of 5G are currently not fully known.

Elon Musk Becomes a Public Figure (2018)

In 2010, Elon Musk was virtually unknown, except to a few electric car converts who saw the future when driving a Tesla Roadster. I've been told many times by viewers of my YouTube channel that they realize, in retrospect, their first time hearing about Musk was in my 2014 Tesla videos. Now, nearing the end of the decade, Elon Musk has become an icon. An engineer with a degree in physics who's famous for building cool stuff: it's not often you see that. Musk is both loved and hated, almost in equal measure. Some love his no-nonsense attitude, but others think he has a big mouth and is a busybody. The negative opinions range from an over-promiser to a downright con artist. Meanwhile, the positive opinions range from a guy who gets stuff done to someone who'll save the planet. In 2018, Musk got into trouble with the Securities and Exchange Commission (SEC) over a tweet. The tweet was an announcement that he would be taking the company private (the legal issue is that the information should not have been made public). He was forced to step down as chairman of the board at Tesla. Ironically, this occurred just as the Tesla Model

3 became the best-selling car in the United States by revenue, and as the excellence of the company's batteries shocked critics in Australia.

Electric Cars Go Mainstream (2018)

Despite early production issues, Tesla's Model 3 was the first "mainstream" production electric car on the road. Tesla had just opened the floodgates. In a few years, traditional companies like Volvo, Jaguar, and Land Rover were vowing to go electric, but only by 2020. Hyundai, Porsche, and Aston Martin also got on the bandwagon. The electric car went from being laughed at, to becoming an expensive dream only for the rich, to trickling down into the mainstream. I had access to a Tesla Model S P100D for a few days in Melbourne in 2017. It was so smooth and quick that it didn't even feel like driving a car. It felt like something else entirely.

Some were worried about replacing batteries in electric cars, but this concern has proven unfounded. In 2018, a Tesla Model S taxi in the Napa Valley crossed the 400,000 km mark on its odometer, with 93 percent battery health remaining at the end.

Lab-Grown Food (2019)

Lab-grown meat was the stuff of science fiction, but in the 2010s it became a reality: Take some stem cells from a host animal, filter them to get muscle and fat tissue that can be grown, and separate the muscle and fat to culture separately. The culturing process uses heat, oxygen, and a serum containing salt, sugar, protein, and nutrients to support cell growth. In this environment, the cells are tricked into thinking they're still inside the host animal and begin to replicate. One stem cell can grow 1 trillion muscle cells. The challenge for scientists is to get the cells to grow in a way that mimics the meat we see in a supermarket. A slab of meat is a complex structure with different kinds of tissue, bone, etc. Currently, it's possible to create uniform mincemeat that looks convincing and even bleeds.

Debuting in 2013, the first lab-grown burger cost $330,000 to grow. The cost was driven by the need for an effective nutrient serum. By 2018, the process had become economical, at $11 per patty (a Big Mac is around $4 on average). To get the cost down further, robotics and machine learning are being used by some companies to sift through a vast array of serum-replacement combinations, to find the cheapest one that works. Other methods include using genetically-engineered plant material.

Arguments have been made for lab meat to feed the growing global population. Compared to other livestock, such as horses, cows need twenty-eight times more land and eleven times more water. Beef production requires 25 percent of global land use and contributes to forestry emissions.

Lab-grown meat is scheduled to be put on the market in 2019.

CHAPTER 15

Looking to the Future

At this point, we've traveled the full distance of our story and become familiar with the new thinkers who built our world. From the simple origins of technology over three hundred years ago, to its growth and explosive maturity today. This book was written in 2018 and new innovations are making headlines every month; it seems fitting to wrap up our journey with a quick look at what could be around the corner for the 2020s.

Household Robots

One of the last remaining problems for robotics is dexterity. That is, handling objects effectively with hand-like appendages. In 2018, the Elon Musk-funded company OpenAI trained a robotic hand to manipulate objects as elegantly as a human. The team used reinforcement learning to achieve this. Reinforcement learning is essentially trial and error. In the experiment, they ran a simulation of the hand manipulating objects in a virtual environment (amounting to one hundred years of experience). The result was the hand manipulating objects it had never seen before, and even mimicking forms of human dexterity (hand sliding and pivoting) without being trained to do so.

If such breakthroughs become integrated with the agility of Boston Dynamics robots, there could be a market for dexterous household robots, or robots for healthcare, within a decade. However, there is some concern about the military implications of this technology.

Battery Tech Improves

As the 2020s roll around, advanced batteries will become the lifeblood of emerging innovation. Already, novel battery technologies are on the horizon. Contenders include:

Gold Nanowire Batteries by the University of California, Irvine

Early tests have completed 200,000 charging cycles in three months, with no signs of degradation.

Graphene Batteries by Samsung and Grabat

Samsung has started working with graphene- (carbon atoms arranged in a sheet one atom's width thick) infused lithium-ion batteries that provide five times faster charging (a smartphone would be charged in twelve minutes) and 45 percent longer battery life.

A company called Grabat has produced graphene batteries that provide 500 miles (804 km) of driving range for electric cars. The batteries can charge a car in a few minutes, 33 times faster than the state of the art.

Dual Carbon Battery by Power Japan Plus

This battery promises sustainability, a twenty-times-faster charge than Samsung's lithium-ion battery, and a duration of 3,000 charge cycles.

Phinergy and Alcoa Canada

A car drove 1,100 miles (1770 km) on a single charge using an oxygen-lithium-ion combination. The test was carried out by Phinergy (an Israeli company) in collaboration with Alcoa Canada.

Whether all these batteries make it to mainstream adoption is beside the point. They show a clear trend: battery technology is yet again advancing. This could solve the major bottleneck in renewable energy, but also lead to new technologies that are currently too expensive or infeasible, like long-distance electric aircraft.

The Web Comes Increasingly Under Threat

As we move toward the 2020s, the open web envisioned by Bush and Lick and implemented by Tim B. Lee is coming under threat. Directives like the link tax from the European Union (a tax on hyperlinks that originate from certain publishers, like news articles, for example) undermine the internet's original purpose.

A letter signed by internet pioneers, including Tim B. Lee, states that, if such a measure as the link tax had been implemented at the internet's creation, "it's unlikely that [the internet] would exist today as we know it. The damage that this may do to the free and open internet as we know it is hard to predict, but in our opinion could be substantial."

The future of the internet is uncertain. I sincerely hope we fight for an open and free internet in the future.

Brain and Technology Merge

Imagine a future where man and machine merge.

Elon Musk's Neural Link aims to do this. Musk stated that we currently have a "slow bandwidth problem" when it comes to interfacing with technology. In other words, using our thumbs and fingers is a slow way to interact with the information we need. The idea behind Neural Link is to create a high-bandwidth "interface to the brain" that would link humans directly with computers, merging us with AI. The brain will act as the limbic system layer (emotions, memories, behavior, stimulation) while the AI acts as another layer to do all the heavy lifting. In theory, anyone who wanted to could achieve vastly superior cognition with a Neural Link installed.

Connecting the human brain directly into technology is a terrifying concept to some, and I would prefer to err on the side of caution for this one.

Alternative Money

As confidence in the current economic system wanes, other forms of money, such as cryptocurrencies, may rise again to prominence. However, this will not be without growing pains. Hacking (possible with future technology) and whether currencies will gain mainstream adoption remain significant issues.

Aside from cryptos, blockchain and other trustless systems will continue to grow in scope and prominence.

The Social Media Backlash

The social media backlash first showed its face in the mid-2010s, led by Facebook. From accusations of causing mental harm to data and security concerns, everybody seems sick of the platform.

In 2018 a Pew Research Center survey showed that 44 percent of young American Facebook users (between eighteen and twenty-nine years of age) had deleted the app from their phones. Facebook might be a leading indicator on the future of social media in general.

I've always wondered why I've felt that social media is a net negative for society. After thinking and researching for some time, here are some reasons why...

Destruction of Human Empathy

Empathy is the ability to feel or understand another person's perspective. Putting yourself in someone else's shoes and basically asking yourself internally, "What if that were me?" Empathy is one of the major elements of humanity that prevent us from displaying animalistic tendencies. Around 2015, studies began to surface finding that social media was biologically damaging the capacity for empathy in its younger users.

The Institute for Social Research found that college students of the late 2010s are 40 percent lower in empathy than were those of the 1990s.

Decline in Mental Health

Most social media platforms, and Facebook in particular, take part in hacking human nature. Engineers at these companies have exploited the biological dopamine reward pathways in our brains to get us to spend more time on the site. For example, likes, comments, and shares release dopamine in our brains, so we try to get a lot of them. Finding a post on Facebook that we enjoy also releases dopamine, so we keep scrolling until we find one.

Unfortunately for some, this can lead to a very real addiction similar to a chemical dependency on drugs. Additionally, we may experience loneliness, anxiety, envy, and even depression when our posts consistently don't get as much interaction as we would like, or when we compare our lives to the perfectly curated, apparent lives seen in others' status updates.

Ex-Facebook president Sean Parker and ex-Facebook Executive Chamath Palihapitiya both came out in 2018 and expressed sorrow and guilt for what they had designed. Palihapitiya stated that social media "is ripping apart society" and "destroying how society works."

Toxic Interactions

As we head toward 2020, we see social media becoming more divisive. People can't seem to talk to each other without getting angry, but why?

Apart from low empathy causing detachment from others' points of view, I feel there might be an additional reason: the Dunning-Kruger effect.

The Dunning-Kruger effect was proposed by David Dunning and Justin Kruger in 1999. They found that, if people have limited knowledge on a topic, they tend to be extremely confident in what they're saying and grossly overestimate their competence to discuss it. Conversely, as people gain more knowledge, they become more shy about expressing it.

If we apply this theory to social media conversations, the people who know the least will be commenting the most because they're over-confident.

The Dunning-Kruger effect, combined with empathy destruction, could create an environment where those with the least knowledge are the most vocal, yet are unable to comprehend opposing points of view. At the same time, those with the most knowledge are likely to stay silent. Hence we end up with a cesspool of over-confident ignoramuses yelling at each other. Social media in a nutshell, ladies and gentlemen.

I think it's safe to say that, if social media giants keep enabling these aspects of human nature, we might see a full-scale social media backlash in the 2020s.

Going to the Moon Again: SpaceX

In 2018, Elon Musk announced that his aerospace company SpaceX (from chapter 13) would be taking eight tourists to the moon in 2023 on a spaceship codenamed the BFR.

The BFR is a massive spacecraft suited for long-distance travel and is capable of carrying 100 people. The mission will launch from Earth and do a loop around the moon before heading back to Earth. The trip is set to take five days in total.

The first passenger for the moon trip will be Japanese billionaire Yusaku Maezawa, who paid for all eight seats. Maezawa is a fan of art and wants to take seven artists (musicians, poets, writers, filmmakers, etc.) with him to see what amazing art they create and share with the rest of humanity.

If all goes well, the first manned trip to orbit the moon will go ahead after numerous unmanned test flights. It seems like a far-out idea, but it'll be amazing to witness if this does become a reality. Optimistically, moon bases and regular civilian moon travel could happen within the decade, with prices falling thereafter. Whether it actually happens or not, it's interesting to think that we live in a time where this sort of vision is feasible.

The Power of New Thinking

In this book, we've seen how our current world was built by those who stood on the shoulders of previous innovators. We can imagine the progress of technology as a massive boulder on flat ground. In the distance is a hill with a steep downward slope. Each innovation pushes the boulder forward a little, and the next innovation pushes a little harder. At first, the boulder only moves slowly, but then it begins to pick up speed. The technological advancement process has been accelerating for hundreds of years. Soon, the boulder reaches the hill with the steep downward slope and rolls faster than anyone could push it. As we approach the year 2020, technological progress is starting to roll down that hill.

The most amazing thing is that a small coterie of key minds made it all possible. These were the inventors—the curious ones who had a new way of thinking. Their minds had the ability to imagine things that didn't exist. They had the gift of pulling abstract ideas from emptiness. Ideas were given form through invention and reason or, curiously, were stumbled upon by accident.

A Chain of Innovation

James Watt's steam engine was the first shove large enough to dislodge the boulder. As it rolled, all other pushes forward in innovation would add to its motion.

Nikola Tesla imagined a world that harnessed electricity. His work, combined with steam, gave manufacturing another push forward. Henry Ford saw how manufacturing could give everyone a car, expanding the very fabric of cities.

Not long afterwards, the Wright brothers gazed at the birds in the air and imagined that man could one day join them. We soon would. Building on their dream, Charles Lindbergh risked his life to fly further than anyone before him. His bravery gave us the air travel industry.

In the midst of turbulent war, the transistor, the lifeblood of the digital age, came about—the result of tireless experimentation and determination from those at Bell Labs.

The boulder representing our progress begins to roll more easily on the
flat ground.

At the war's end, Vannevar Bush declares that leading scientists should turn their minds away from destruction and toward inventions that would benefit all. In a visionary piece, he imagines the Memex: a machine that could store, search, and link all information to be shared by everyone.

Joseph Licklider and Ivan Sutherland become inspired by Bush's new thinking. Licklider uses the idea to create what later becomes the internet. Ivan Sutherland also takes the torch of innovation from Bush in Sketchpad. Sutherland imagines that humans and computers could work together in real time. This is the start of computer graphics.

The torch of Bush is passed from Sutherland to Douglas Engelbart. Engelbart takes computer graphics to a point where humans and computers can interact freely, without punch cards. Engelbart's ideas of a mouse, keyboard, graphics, and networked computers birth the concept of the desktop work station.

Meanwhile, the dream team of scientists from Fairchild turn transistors into the CPU—the brain of computers. From this moment, the acceleration of computing power can support the next wave of new ideas. Soon, Bill Gates opens up computer coding to everyone.

At this point, the boulder is starting to noticeably accelerate along the *flat ground.*

Engelbart's new thinking encourages innovation within the very Xerox PARC team that inspired Steve Jobs. Jobs then brings graphical computers to the world with the Mac.

Engelbart's core idea of making human-computer interaction more intuitive flows through to Steve Jobs' NeXTSTEP software, which is used to build the World Wide Web and eventually comprises the seeds for the iPhone. Both of these change the world forever.

By now, the boulder is moving at blistering speed, though it is still on flat *ground. However, the downhill slope is only seconds away.*

The World Wide Web builds on top of the internet and joins together the minds of brilliant people around the world. Knowledge is available and shared like never before, just as Bush had imagined. Then suddenly—just like that, it seems—we have stem cells that can regenerate organs, DNA editing, lab-grown food, athletic robots, electric cars, solar planes, new battery technology, artificial intelligence making sense of vast seas of data, quantum computers promising unimaginable power...

The boulder is rolling down the hill toward top speed, traveling faster than we could ever push it.

It seems like great minds have stopped pushing, and like gravity, their inventions have taken over the downhill motion.

But what happens next? Does the boulder rolling by itself mean that we will reach a utopia where we no longer work? Are humans finally free from drudgery? Or does the boulder in free-fall signal that things are out of our control?

We've only just reached the hill, so there's only one thing that we can say for sure: The story of our journey to this point, and the new thinking that brought us here, is one of the greatest stories in existence—a story that is still continuing.

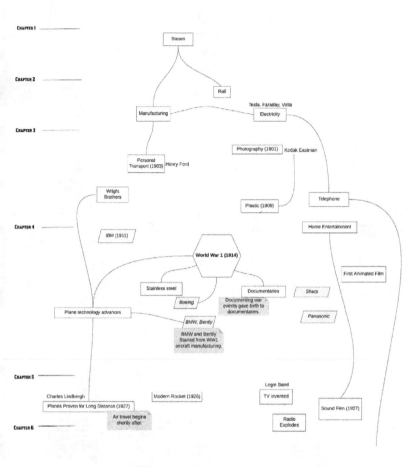

Chapter 1

Steam

Chapter 2

Rail

Manufacturing

Tesla, Faraday, Volta
Electricity

Chapter 3

Photography (1901)   Kodak Eastman

Personal
Transport (1903)   Henry Ford

Wright
Brothers

Plastic (1909)

Telephone

Chapter 4

IBM (1911)

Home Entertainment

World War 1 (1914)

First Animated Film

Stainless steel

Documentaries   Sharp

Boeing

Documenting war
events gave birth to
documentaries.

Panasonic

Plane technology advances

BMW, Bently

BMW and Bently
Started from WW1
aircraft manufacturing.

Chapter 5

Logie Baird

Charles Lindbergh   Modern Rocket (1926)   TV invented

Planes Proven for Long Distance (1927)

Air travel begins
shortly after.

Radio
Explodes

Sound Film (1927)

Chapter 6

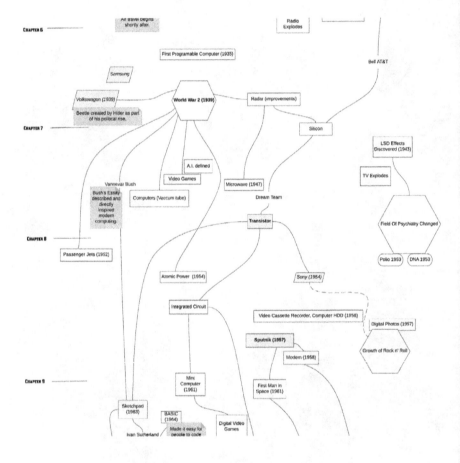

Chapter 6

Air travel begins shortly after.

Radio Explodes

First Programable Computer (1935)

Bell AT&T

Samsung

Volkswagon (1939)

World War 2 (1939)

Radar (improvements)

Beetle created by Hitler as part of his political rise.

Chapter 7

Silicon

LSD Effects Discovered (1943)

A.I. defined

Video Games

TV Explodes

Vannevar Bush

Microwave (1947)

Computers (Vaccum tube)

Dream Team

Bush's Essay described and directly inspired modern computing.

Field Of Psychiatry Changed

Transistor

Chapter 8

Polio 1953   DNA 1953

Passenger Jets (1952)

Atomic Power (1954)

Sony (1954)

Integrated Circuit

Video Cassette Recorder, Computer HDD (1956)

Digital Photos (1957)

Sputnik (1957)

Growth of Rock n' Roll

Modem (1958)

Chapter 9

Mini Computer (1961)

First Man in Space (1961)

Sketchpad (1963)

BASIC (1964)

Digital Video Games

Ivan Sutherland

Made it easy for people to code

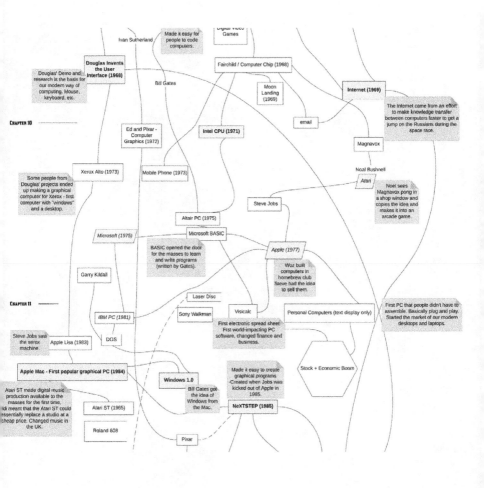

Ivan Sutherland

Made it easy for people to code computers.

Digital Video Games

Douglas Invents the User Interface (1968)

Douglas' Demo and research is the basis for our modern way of computing. Mouse, keyboard, etc.

Bill Gates

Fairchild / Computer Chip (1968)

Moon Landing (1969)

Internet (1969)

The internet came from an effort to make knowledge transfer between computers faster to get a jump on the Russians during the space race.

CHAPTER 10

Ed and Pixar - Computer Graphics (1972)

email

Intel CPU (1971)

Magnavox

Xerox Alto (1973)

Mobile Phone (1973)

Noal Bushnell

Atari

Noel sees Magnavox pong in a shop window and copies the idea and makes it into an arcade game.

Some people from Douglas' projects ended up making a graphical computer for Xerox - first computer with "windows" and a desktop.

Steve Jobs

Altair PC (1975)

Microsoft (1975)

Microsoft BASIC

Apple (1977)

BASIC opened the door for the masses to learn and write programs (written by Gates).

Woz built computers in homebrew club Steve had the idea to sell them.

Garry Kildall

CHAPTER 11

Laser Disc

First PC that people didn't have to assemble. Basically plug and play. Started the market of our modern desktops and laptops.

IBM PC (1981)

Sony Walkman

Visicalc

Personal Computers (text display only)

First electronic spread sheet First world-impacting PC software, changed finance and business.

Steve Jobs saw the xerox machine.

Apple Lisa (1983)

DOS

Apple Mac - First popular graphical PC (1984)

Atari ST made digital music production available to the masses for the first time, ulid meant that the Atari ST could essentially replace a studio at a cheap price. Changed music in the UK.

Windows 1.0

Bill Gates got the idea of Windows from the Mac.

Made it easy to create graphical programs -Created when Jobs was kicked out of Apple in 1985.

Stock + Economic Boom

Atari ST (1985)

NeXTSTEP (1985)

Roland 808

Pixar

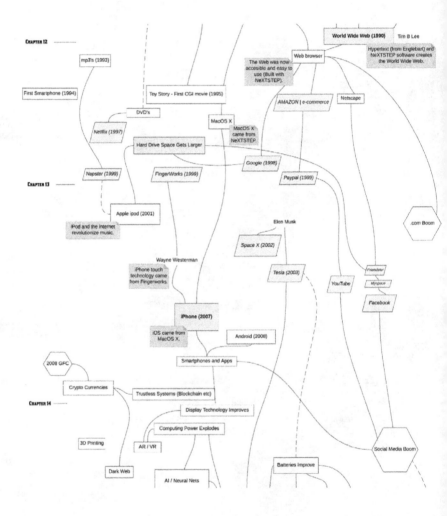

CHAPTER 12

World Wide Web (1990)   Tim B Lee

mp3's (1993)

Hypertext (from Englebart) and NeXTSTEP software creates the World Wide Web.

Web browser

The Web was now accesible and easy to use (Built with NeXTSTEP).

First Smartphone (1994)

Toy Story - First CGI movie (1995)

AMAZON | e-commerce   Netscape

DVD's

MacOS X

MacOS X came from NeXTSTEP

Netflix (1997)

Hard Drive Space Gets Larger

Google (1998)

CHAPTER 13

Napster (1999)   FingerWorks (1999)   Paypal (1999)

Apple ipod (2001)

.com Boom

iPod and the internet revolutionize music.

Elon Musk

Space X (2002)

Wayne Westerman

iPhone touch technology came from Fingerworks.

Tesla (2003)

Friendster

YouTube   Myspace

Facebook

iPhone (2007)

iOS came from MacOS X.

Android (2008)

2008 GFC

Smartphones and Apps

Crypto Currencies

CHAPTER 14

Trustless Systems (Blockchain etc)

Display Technology Improves

Computing Power Explodes

3D Printing

AR / VR

Social Media Boom

Dark Web

Batteries Improve

AI / Neural Nets

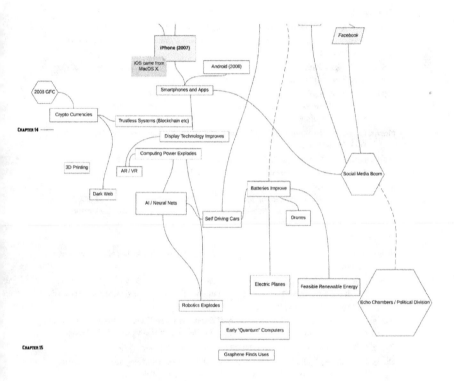

iPhone (2007)

iOS came from
MacOS X.

Android (2008)

Smartphones and Apps

2008 GFC

Crypto Currencies

Trustless Systems (Blockchain etc)

Chapter 14

Display Technology Improves

Computing Power Explodes

3D Printing

AR / VR

Dark Web

AI / Neural Nets

Batteries Improve

Self Driving Cars

Drones

Facebook

Social Media Boom

Electric Planes

Feasible Renewable Energy

Robotics Explodes

Echo Chambers / Political Division

Early "Quantum" Computers

Chapter 15

Graphene Finds Uses

Bibliography

New Thinking is a work of nonfiction. It is not intended to be a replacement, replication of academic text nor a peer-reviewed document. The author cites the following resources as guides into the book and as future readings for a wider understanding.

Chapter 1

California State University, Northridge, "The Impact of the Railroad: The Iron Horse and the Octopus," accessed November 27, 2018, http://www.csun.edu/~sg4002/courses/417/readings/rail.pdf.

Lira, Carl, Michigan State University, "Biography of James Watt," last modified May 21, 2013, https://www.egr.msu.edu/~lira/supp/steam/wattbio.html.

McNeil, Ian, ed., An Encyclopaedia of the History of Technology (New York: Routledge, 1990). Accessed November 28, 2018, https://polifilosofie.files.wordpress.com/2012/12/encyclopedia-of-the-history-of-technology.pdf.

Morris, Charles R. The Dawn of Innovation: The First American Industrial Revolution 1st ed. (New York: PublicAffairs, 2012).

Stanford University Engineering, "History of American Railroads," accessed November 28, 2018, https://cs.stanford.edu/people/eroberts/cs181/projects/corporate-monopolies/development_rr.html.

The Telegraph, "The power behind the Industrial Revolution," last modified June 22, 2000, https://www.telegraph.co.uk/news/science/science-news/4750891/The-power-behind-the-Industrial-Revolution.html.

Chapter 2

Bellis, Mary, Timeline: Biography of Samuel Morse 1791–1872 (New York: The New York Times Company, 2009).

Grimnes, Sverre and Orjan G. Martinsen, Bioimpedance & Bioelectricity Basics 2nd ed. (Oxford: Elsevier, 2008), https://books.google.com.au/books?id=v3EuUjoqwkkC&pg=PA411&redir_esc=y#v=onepage&q&f=true.

McNeil, Ian, ed., An Encyclopaedia of the History of Technology (New York: Routledge, 1990). https://polifilosofie.files.wordpress.com/2012/12/encyclopedia-of-the-history-of-technology.pdf.

Petrie, A. Roy, Alexander Graham Bell (Ontario: Fitzhenry & Whiteside, 1975).

University of Salford, "John Tawell: The Man Hanged by the Electric Telegraph," last accessed November 28, 2018, http://www.cntr.salford.ac.uk/comms/johntawell.php.

West, Krista, The Basics of Metals and Metalloids (New York: Rosen Publishing Group, 2013).

Chapter 3

Anderson, John David, Introduction to Flight (New York: McGraw-Hill Education, 2004).

Bryan, Ford, Henry Ford Heritage Association, "The Birth of Ford Motor Company," accessed November 28, 2018, http://hfha.org/the-ford-story/the-birth-of-ford-motor-company/.

Camera of the Month, "Kodak Brownie Target Six-20," last updated July 1, 2001, http://www.cameraofthemonth.com/articles/KodakBrownie.shtml.

Daimler, "The history behind the Mercedes-Benz brand and the three-pointed star," last updated June 5, 2007, https://media.daimler.com/marsMediaSite/en/instance/ko/The-history-behind-the-Mercedes-Benz-brand-and-the-three-pointed-star.xhtml?oid=9912871.

DeVinney, James A., dir. "George Eastman: The Wizard of Photography," *The American Experience Series*, PBS, released 2000.

Emspak, Jesse, Live Science, "8 Ways You Can See Einstein's Theory of Relativity in Real Life," last modified March 14, 2017, https://www.livescience.com/58245-theory-of-relativity-in-real-life.html.

Flyingmachines.org, "Clement Ader," accessed November 28, 2018, http://www.flyingmachines.org/jatho.html.

Flyingmachines.org, "Karl Jatho," accessed November 28, 2018, http://www.flyingmachines.org/jatho.html.

Ford, Henry, *My Life and Work*. Greenbook Publications, LLC. August 1, 2010.

Goss, Jennifer L., ThoughtCo., "Henry Ford and the Auto Assembly Line," last modified January 23, 2018, https://www.thoughtco.com/henry-ford-and-the-assembly-line-1779201.

Gordon, Robert J., *The Rise and Fall of American Growth*, (Princeton: Princeton University Press, 2016). https://books.google.com.au/books?id=bA8mDwAAQBAJ&pg=PA188&lpg=PA188&dqe#v=onepage&q&f=false.

History, "Ford's assembly line starts rolling," last modified August 21, 2018, https://www.history.com/this-day-in-history/fords-assembly-line-starts-rolling.

Kenney, William Howland, *Recorded Music in American Life: The Phonograph and Popular Memory, 1890–1945*, (Oxford: Oxford University Press, 1999).

Kettering, Charles Franklin (1946). Biographical memoir of Leo Hendrik Baekeland, 1863-1944. Presented to the academy at the autumn meeting, 1946. National Academy of Sciences (U.S.).

Linehan, Andrew. "Soundcarrier." Continuum Encyclopedia of Popular Music of the World. pp. 359–366."

Smithsonian National Air and Space Museum, "The 1902 Glider," accessed November 28, 2018, https://airandspace.si.edu/exhibitions/wright-brothers/online/fly/1902/glider.cfm.

The Museum of Unnatural History, "Gustave Whitehead: Did He Beat the Wright Brothers into the Sky?" accessed November 28, 2018, http://unmuseum.mus.pa.us/gustave.htm.

"The Wright Stuff: The Wright Brothers and the Invention of the Airplane," *The American Experience Series*, PBS, released 1996.

Chapter 4

Barnard, Timothy and Peter Rist, *South American Cinema: A Critical Filmography, 1915–1994*, (New York: Routledge, 2011).

Boeing, "Boeing History," accessed November 28, 2018, http://www.boeing.com/history/.

Encyclopeida.com, "Boeing Company," accessed November 28, 2018, http://www.encyclopedia.com/history/united-states-and-canada/us-history/boeing-company.

Hart, Peter, *The Great War: 1914–1918*, (London: Profile Books, 2013).

IBM, "Chronological History of IBM," last accessed November 28, 2018, https://www-03.ibm.com/ibm/history/history/decade_1910.html.

Kelly, Martin, ThoughtCo., "5 Key Causes of World War I," accessed November 28, 2018, https://www.thoughtco.com/causes-that-led-to-world-war-i-105515.

MacDonald, Lyn, *To the Last Man: Spring 1918*, (Eastbourne, Gardners Books, 1999).

Noakes, Andrew, *The Ultimate History of BMW*, (Bath: Parragon Publishing, 2005).

Sass, Erik, Mental Floss, "12 Technological Advances of World War I," last modified April 30, 2017, http://mentalfloss.com/article/31882/12-technological-advancements-world-war-i.

Sass, Erik, Mental Floss, "WWI Centennial: The Second Bolshevik Coup Attempt Succeeds," last modified November 8, 2017, http://mentalfloss.com/article/514164/wwi-centennial-second-bolshevik-coup-attempt-succeeds.

Sharp World, "A Century of Sincerity and Creativity," last accessed November 28, 2018, http://www.sharp-world.com/corporate/img/info/his/h_company/pdf_en/all.pdf.

Sheffield, Gary, *A Short History of the First World War*, (London, Oneworld Publications, 2014).

Strachan, Hew, *The First World War*, (New York: Simon & Schuster, 2014).

Strite, Charles. Bread-toaster. US Patent #1,387,670, filed May 29, 1919, and issued August 16, 1921.

Taylor, A J P, *The First World War: An Illustrated History*, (New York: The Berkeley Publishing Group, 1972).

The New York Times, "A Non-Rusting Steel: Sheffield Invention Especially Good for Table Cutlery," last updated January 31, 1915, https://www.nytimes.com/1915/01/31/archives/a-nonrusting-steel-sheffield-invention-especially-good-for-table.html.

Weiss, Stanley I. and Amir R. Amir, Encyclopaedia Britannica, "Boeing Company," accessed November 28, 2018, https://www.britannica.com/topic/Boeing-Company.

Chapter 5

Abramson, Albert, *The History of Television, 1880 to 1941*, (Jefferson: McFarland, 1987).

Burns, R. W., *Television: An International History of the Formative Years*, (London: The Institute of Engineering and Technology, 1998).

Craig, Douglas B., *Fireside Politics: Radio and Political Culture in the United States, 1920–1940*, (Baltimore: The Johns Hopkins University Press, 2000).

Fellow, Anthony, *American Media History*, 3rd ed. (Boston: Cengage Learning, 2012).

Goddard, Robert, *Rocket Development: Diary of the Space Age Pioneer*, (New York: Prentice-Hall, 1961).

Gugliotta, Guy, *Discover*, "How Radio Changed Everything," last modified May 31, 2007, http://discovermagazine.com/2007/jun/tireless-wireless.

Hutchinson, Ron, "The Vitaphone Project. Answering Harry Warner's Question: 'Who the Hell Wants to Hear Actors Talk?'" *Film History* Vol 1, No 1 (2002) 40–46. Accessed November 28, 2018, https://www.jstor.org/stable/3815579.

InflationData.com, "Inflation and CPI Consumer Price Index 1920-1929," accessed November 28, 2018, https://inflationdata.com/articles/inflation-consumer-price-index-decade-commentary/inflation-cpi-consumer-price-index-1920-1929/.

Kurzweil, Ray, *The Singularity is Near*, (New York: Penguin Group, 2005).

Lehman, Milton, *This High Man: The Life of Robert H. Goddard*, (New York: Farrar, Strauss and Co., 1963).

"Lost Peace," *People's Century 1900–1999*, PBS/BBC, released 1998.

NASA "Dr. Robert H. Goddard: American Rocketry Pioneer," accessed November 28, 2018, https://www.nasa.gov/centers/goddard/about/history/dr_goddard.html

"On the Line," *People's Century 1900–1999*, PBS/BBC, released 1998.

Pandora Archive. Accessed November 28, 2018, Pandora.nla.gov.au.

Project Twenty: The Jazz Age, Shanchie Records, released 2003.

Sann, Paul, "The Lawless Decade," accessed November 28, 2018, http://www.lawlessdecade.net/.

Science Stuck, "41 Ingeniously Smart Inventions of the 1920s You Should Know About," accessed November 28, 2018, https://www.buzzle.com/articles/inventions-of-the-1920s.html.

Segal, David, *Materials for the 21st Century*, (Oxford: Oxford University Press, 2017).

Stefanyshyn, Deanna and Julie Kendell, *ETEC540: Text, Technologies—Community Weblog*, "The Influence of Radio and Television on Culture, Literacy and Education," last updated October 28, 2012, http://blogs.ubc.ca/etec540sept12/2012/10/28/1687/.

"The Story of Charles A. Lindbergh," *Famous Americans of the 20th Century*, Questar, released 1991.

ThoughtCo. "20th Century Invention Timeline: 1900 to 1949," accessed November 28, 2018, https://www.thoughtco.com/20th-century-timeline-1992486.

University of Minnesota Libraries, "7.4 Radio's Impact on Culture," accessed November 28, 2018, http://open.lib.umn.edu/mediaandculture/chapter/7-4-radios-impact-on-culture/.

Chapter 6

Barber, Chris, *Birth of the Beetle: The Development of the Volkswagen by Ferdinand Porsche*, (Sparkford: Haynes Publishing, 2003).

Blecha, Peter, *Vintage Guitar* magazine "Audiovox #736: The World's First Electric Bass Guitar!" last modified 1999, https://www.vintageguitar.com/1782/audiovox-736/.

Edward, David and Mike Callahan, Patrice Eyries, Randy Watts, and Tim Neely, Both Sides Now Publications, "RCA Program Transcription Album Discography (1931–33)," accessed November 28, 2018, http://bsnpubs.com/rca/rca/rca33.html.

Encyclopedia.com, "Samsung Group," accessed November 28, 2018, http://www.encyclopedia.com/books/politics-and-business-magazines/samsung-group.

Encyclopedia.com, "Turing Machine," accessed November 28, 2018, http://www.encyclopedia.com/science-and-technology/computers-and-electrical-

Encyclopedia.com, "Volkswagen Beetle," accessed November 28, 2018, http://www.encyclopedia.com/history/culture-magazines/volkswagen-beetle.

Frank, Robert H. and Ben S. Bernanke, *Principles of Macroeconomics* 3rd ed., (Boston: McGraw-Hill/Irwin, 2007).

Garraty, John A., *The Great Depression* (New York: Anchor, 1987)

Hoffman, David, *How Hitler Lost the War*, PBS, released 2012.

IDSIA Dalle Molle Institute for Artificial Intelligence, "Konrad Zuse (1910–1995), accessed November 28, 2018, http://people.idsia.ch/~juergen/zuse.html.

InflationData.com, "Inflation and CPI Consumer Price Index 1930-1939," accessed November 28, 2018, https://inflationdata.com/articles/inflation-consumer-price-index-decade-commentary/inflation-cpi-consumer-price-index-1930-1939/.

I-Programmer, "Konrad Zuse and the First Working Computer," last modified June 19, 2013, http://www.i-programmer.info/history/people/253-konrad-zuse.html.

Isaacson, Walter, *The Innovators: How a Group of Inventors, Hackers, Geniuses and Geeks Created the Digital Revolution*, (New York: Simon & Schuster Paperbacks, 2014).

King, Susan, *Los Angeles Times* "How did 'Wizard of Oz' fare on its 1939 release?" last modified March 11, 2013, http://articles.latimes.com/2013/mar/11/entertainment/la-et-mn-original-wizard-reaction-20130311.

Krugman, Paul, *New York Times* "Protectionism and the Great Depression," last modified November 30, 2009, https://krugman.blogs.nytimes.com/2009/11/30/protectionism-and-the-great-depression/.

Laidler, David E.W. and George W. Stadler, "Monetary Explanations of the Weimar Republic's Hyperinflation: Some Neglected Contributions in Contemporary German Literature," *Journal of Money, Credit and Banking*, Vol 30, 816–818.

Schultz, Stanley K. *Crashing Hopes: The Great Depression*. American History 102: Civil War to the Present. University of Wisconsin–Madison. (1999).

The Economist "The Great Depression," accessed November 28, 2018, https://www.economist.com/topics/great-depression.

Torchinsky, Jason, Jalopnik, "The Real Story behind the Nazis and Volkswagen," last modified October 2, 2015, https://jalopnik.com/the-real-story-behind-the-nazis-and-volkswagen-1733943186.

Willson, Quentin, *The Ultimate Classic Car Book*, (New York: DK Publishing, 1995).

Wired, "The Decades that Invented the Future, Part 4: 1931–1940," last modified November 9, 2012, https://www.wired.com/2012/11/the-decades-that-invented-the-future-part-4-1931-1940/.

Chapter 7

American Institute of Physics, "Center for History of Physics," Accessed November 28, 2018, https://www.aip.org/.

Axelrod, Alan, *Encyclopedia of World War II*, Vol 1, (New York: Facts on File, 2007).

Bean-Mellinger, Barbara, Bizfluent, "Major Inventions in the 1940s," last modified October 25, 2018, https://bizfluent.com/info-8167994-major-inventions-1940s.html.

Computer History Museum, "The Atanasoff-Berry Computer In Operation" YouTube, accessed November 28, 2018, released July 2010, https://www.youtube.com/watch?v=YyxGlbtMS9E.

Donald, David ed., *Warplanes of the Luftwaffe*, (New York: Aerospace Publishing, 1997).

Eckert Jr, John Presper, and John W. Mauchly. Electronic Numerical Integrator and Computer. US Patent 3,120,606, filed June 26, 1947.

Engineering and Technology History Wiki, "Milestones: Atanasoff-Berry Computer, 1939," last modified December 31, 2018, https://ethw.org/Milestones:Atanasoff-Berry_Computer,_1939.

Guarnieri, Massimo, "Seventy Years of Getting Transistorized,." *IEEE Industrial Electronics Magazine* Vol 11, No 2 (December 2017) 33–37. Accessed November 28, 2018, https://www.researchgate.net/publication/321991346_Seventy_Years_of_Getting_Transistorized_Historical

Keegan, John, *The Second World War*, (London: Pimlico, 1997).

Light, Jennifer S., "When Computers Were Women," *Technology and Culture* Vol 40, No 3 (July 1999) 455–483. Accessed November 28, 2018, https://www.jstor.org/stable/25147356?seq=1#page_scan_tab_contents.

Moye, William T., US Army Research Laboratory, "ENIAC: The Army-Sponsored Revolution," last modified January 1996, http://ftp.arl.mil/mike/comphist/96summary/index.html.

PBS, "Transistorized!" Accessed November 28, 2018, http://www.pbs.org/transistor/album1/.

PBS, *Transistorized!*, YouTube, accessed November 28, 2018, released October 2014, https://www.youtube.com/watch?v=U4XknGqr3Bo.

Simpson, Andrew, Royal Air Force Museum, "Individual History," accessed November 28, 2018, https://www.rafmuseum.org.uk/documents/collections/85-AF-66-Me-163B.pdf.

Tyner, James A. *War, Violence, and Population: Making the Body Count*, (New York: The Guilford Press, 2009).

Chapter 8

Brookhaven National Library. "A Passion for Discovery, a History of Scientific Achievement" http://www.bnl.gov/bnlweb/history/higinbotham.asp

CyberneticZoo.com. "1954— Programmed Article Transfer Patent—George C. Devol Jr. (American)." http://cyberneticzoo.com/early-industrial-robots/1954-programmed-article-transfer-patent-george-c-devol-jr-american/.

Flatow, Ira. *They All Laughed... From Light Bulbs to Lasers: The Fascinating Stories Behind the Great Inventions That Have Changed Our Lives.* HarperCollins, 1993.

Hamborsky J, Kroger A, Wolfe C, eds. (2015), "Poliomyelitis," Epidemiology and Prevention of Vaccine-Preventable Diseases (The Pink Book) (13th ed.), Washington DC: Public Health Foundation, "Poliomyelitis Fact sheet No. 114."

History.com Editors. "Salk announces polio vaccine." HISTORY. February 9, 2010. https://www.history.com/this-day-in-history/salk-announces-polio-vaccine

Kallen, Stuart (1999). A Cultural History of the United States. San Diego: Lucent.

Lemelson-MIT. "Charles Ginsburg: Videotape Recorder (VTR)." https://lemelson.mit.edu/
resources/charles-ginsburg

Living History Farm. "Television" https://livinghistoryfarm.org/farminginthe50s/life_17.html

Malone, Bob. "George Devol: A Life Devoted to Invention, and Robots." Spectrum. September
26, 2011. https://spectrum.ieee.org/automaton/robotics/industrial-robots/george-devol-a-
life-devoted-to-invention-and-robots

Owens, Jeff. "The History of the Fender Stratocaster: The 1950s." Fender. https://www.fender.
com/articles/gear/the-history-of-the-fender-stratocaster-the-1950s

PBS. "Integrated Circuits: 1958: Invention of the Integrated Circuit." http://www.pbs.org/
transistor/background1/events/icinv.html

PBS. "The Discovery of DNA's Structure" http://www.pbs.org/wgbh/evolution/
library/06/3/l_063_01.html

Perales, Marian; Burns, Vincent. Crew, Spencer R.; Mullin, Chris; Swanson, Krister. "American
History." ABC-CLIO, 2012. 11 Dec. 2012.

Profiles in Science. "The Francis Crick Papers: The Discovery of the Double Helix, 1951-1953."
https://profiles.nlm.nih.gov/SC/Views/Exhibit/narrative/doublehelix.html

Puiu, Tibi. "Your smartphone is millions of times more powerful than all of NASA's combined
computing in 1969." ZME Science. Sept 10, 2017. https://www.zmescience.com/research/
technology/smartphone-power-compared-to-apollo-432/.

Robotics.org. "UNIMATE: The First Industrial Robot." https://www.robotics.org/joseph-
engelberger/unimate.cfm

Stevenson, N.J. *Fashion: A Visual History from Regency & Romance to Retro & Revolution.* 2012. New
York City: St. Martin's Griffin.

U-Creative. "Marketing Meets Industrial Design: The Fender Stratocaster Headstock Story."
December 24, 2015. http://www.ucreative.com/articles/fender-strat-headstock/.

Vibrationdata.com. "Sputnik." https://www.vibrationdata.com/Sputnik.htm

Wall, Mike. "Sputnik 1! 7 Fun Facts About Humanity's First Satellite." Space.com. October 2,
2017. https://www.space.com/38331-sputnik-satellite-fun-facts.html

InflationData.com. "Inflation and CPI Consumer Price Index 1950–1959." https://inflationdata.
com/articles/inflation-cpi-consumer-price-index-1950-1959/.

Pushkar, Roberto G. "Comet's Tale." *Smithsonian.com.* June 2002. https://www.smithsonianmag.
com/history/comets-tale-63573615/.

APS Physics. "This Month in Physics History." October 2008. https://www.aps.org/publications/
apsnews/200810/physicshistory.cfm

O'Neil, Mark. "The History of the Modem." MUO. November 28, 2011. https://www.makeuseof.
com/tag/infographic-history-modem/.

Biography.com. "Rosalind Franklin Biography." April 2014. https://www.biography.com/people/
rosalind-franklin-9301344

World Health Organization. "Poliomyelitis" March 14, 2018. http://www.who.int/mediacentre/
factsheets/fs114/en/.

Bellis, Mary. "History of the Medium." *ThoughtCo.* December 31, 2017. https://www.thoughtco.
com/history-of-the-modem-4077013

Chapter 9

Bill & Melinda Gates Foundation. "Bill Gates, 2007 Harvard Commencement." Press Room,
Speeches. June 6, 2007. https://www.gatesfoundation.org/media-center/speeches/2007/06/
bill-gates-harvard-commencement.

Baker, Jeff. "All times a great artist, Ken Kesey is dead at age 66." *The Oregonian.* November 11,
2001.

Bilgil, Melih. *History of the Internet.* http://www.lonja.de

Blitz, Matt. "The Mysterious Death of the First Man in Space." *Popular Mechanics.* April 12, 2016. https://www.popularmechanics.com/space/a20350/yuri-gagarin-death/.

Bush, Vannevar. "As We May Think." *The Atlantic.* July 1945. https://www.theatlantic.com/magazine/archive/1945/07/as-we-may-think/303881/.

Computer History Museum. *Intel 4004 Microprocessor 35th Anniversary.* November 2007.

Computer History Museum. *The Story of the Intel 4004.* 2011

Computer History. "Internet History of 1960s." http://www.computerhistory.org/internethistory/1960s/.

Cringely, Robert X. "Glory of the Geeks," 1998, *PBS.*

Dalakov, Georgi. "The MEMEX of Vannevar Bush." History of Computers. http://history-computer.com/Internet/Dreamers/Bush.html

Doug Engelbart Institute. "About NLS/Augment." https://www.dougengelbart.org/about/augment.html

Faggin, Federico. "The New Methodology for Random Logic Design." Intel 4004.

Frum, David. *How We Got Here: The '70s.* New York: Basic Books, 2000.

Gates, Bill. *The Road Ahead.* 1996. Penguin Books.

Hofmann, Albert, "From Remedy to Inebriant." *LSD: My Problem Child.* New York: McGraw-Hill. 1980. p. 29. ISBN 978-0-07-029325-0.

Hofmann, Albert; translated from the original German *LSD Ganz Persönlich* by J. Ott. MAPS-Volume 6, Number 69, Summer 1969.

Houston, Ronald D.; Harmon, Glynn. "Vannevar Bush and memex." 2007. Annual Review of Information Science and Technology.

McCracken, Harry. "Fifty Years of BASIC, the Programming Language That Made Computers Personal." *Time.* April 29, 2014. http://time.com/69316/basic/.

McMahon, Tim. "Inflation and CPI Consumer Price Index 1970–1979." *InflationData.com.*

National Inventors Hall of Fame. "Marcian E. (Ted) Hoff." https://www.invent.org/inductees/marcian-e-ted-hoff

Nichols, David E. "Serotonin, and the Past and Future of LSD." 2013. *MAPS Bulletin.* https://www.maps.org/news-letters/v23n1/v23n1_p20-23.pdf

Picolsigns. "History of the Internet." YouTube. https://www.youtube.com/watch?v=9hIQjrMHTv4

Reinhold, Robert. "Dr. Vannevar Bush Is Dead at 84." *New York Times.* June 30, 1974. https://www.nytimes.com/1974/06/30/archives/dr-vannevar-bush-is-dead-at-84-dr-vannevar-bush-who-marshaled.html

Saunacy, Ferro. "USPS's 'Informed Delivery' Service Will Email You Pictures of the Mail You're Getting Today," April 6, 2018. *Mental Floss.* http://mentalfloss.com/article/539095/uspss-informed-delivery-service-will-email-you-pictures-mail-youre-getting-today

Silicon Valley, PBS, February 5, 2013.

The 8-Bit Guy. "The basics of BASIC, the programming language of the 1980s." https://www.youtube.com/watch?v=seM9SqTsRG4

Wikipedia. "Licklider, J. C. R.." https://en.wikipedia.org/wiki/J._C._R._Licklider

Wolfe, Tom. "The 'Me' Decade and the Third Great Awakening," August 23, 1976. *New York* magazine.

Zachary, G. Pascal. *Endless Frontier: Vannevar Bush, Engineer of the American Century.* 1997. New York: The Free Press. ISBN 0-684-82821-9. OCLC 36521020.

Chapter 10

Brownlee, Christen, "Biography of Rudolf Jaenisch" 101(39): 13982–13984. *PNAS.* Sep 21, 2004. https://www.ncbi.nlm.nih.gov/pmc/articles/PMC521108/.

Dormehl, Luke. "Today in Apple history: Homebrew Computer Club meets for first time." March 3, 2017. Cult of Mac. https://www.cultofmac.com/470195/apple-history-homebrew-computer-club/.

Haire, Meaghan. "The Walkman," July 1, 2009 *Time*. ISSN 0040-781X.

Hormby, Thomas. "VisiCalc and the Rise of the Apple II." September 25, 2006. *Low End Mac*. https://lowendmac.com/2006/visicalc-and-the-rise-of-the-apple-ii/.

"Microsoft history." The History of Computing Project.

Motherboard. "Meet the Inventor of the First Cell Phone," June 24, 2015. YouTube. https://www.youtube.com/watch?v=C6gNeKjC9Cc.

Nightline. "1981 Nightline interview with Steve Jobs." https://www.youtube.com/watch?v=3H-Y-D3-j-M.

PC History. "How the Altair Began." http://www.pc-history.org/altair.htm.

Sony Soundabout, Stowaway, Freestyle. https://www.sony.co.jp/SonyInfo/CorporateInfo/History/capsule/20/.

Talk of the Nation. "A Chat with Computing Pioneer Steve Wozniak." September 29, 2006. NPR. https://www.npr.org/templates/story/story.php?storyId=6167297.

The Sydney Morning Herald. "Apple co-founder tells his side of the story." Fairfax Media. September 28, 2006.

Triumph of the Nerds, 1996, PBS, Robert X. Cringely.

Steve Jobs: The Lost Interview, 2012, Robert X. Cringely.

Chapter 11

Bunnell, David "The Man Behind The Machine?" *PC Magazine* (interview), 1982

G4 Icons. "History of the NES." Season 4 Episode 10, 2005

Koblin, John "MTV Mines the Past for Its Future: 'Total Request Live.'" *New York Times*, 2017

Sandberg-Diment, Erik "Personal Computers; Hardware Review: Apple Weighs in with Macintosh." 1984

Sirota, David "Back to Our Future: How the 1980s Explain the World We Live in Now" *Our Culture, Our Politics, Our Everything*, 2011

Chapter 12

Ashcraft, Brian. "What's The Father of the PlayStation Doing These Days?" *Kotaku*, 2010.

Businessinsider.com. "What is the internet and how the internet works." 2011. http://www.businessinsider.com/what-is-the-internet-and-how-the-internet-works2011-6?IR=T#now-heres-how-it-works-from-the-moment-you-sign-online-your-computer-becomes-one-of-the-millions-of-machines-contributing-to-the-internet-4

Edge Staff. "The Making Of: PlayStation." Edge Online. *Edge Magazine*.2009.

Internetlivesstats.com. "Total number of websites." 2018. http://www.internetlivestats.com/total-number-of-websites/.

Mixergy. "The Story Behind Pixar, with Alvy Ray Smith." Alvy Ray Smith, Archived from the original 2005.

Mobvity.com. "The First Text Message" (1992). https://www.mobivity.com/2012/09/a-brief-history-of-text-messaging/.

Ohio State University. "Pixar Animation Studios." Retrieved April 22, 2008.

Price, David A. "The Pixar Touch: The making of a Company (1st ed.)." *New York: Alfred A. Knopf*, 2005

Smith, Alvy Ray. "*Pixar Founding Documents*." Alvy Ray Smith, Archived from the original 2005.

Statista.com. "Number of internet user worldwide." 2018. https://www.statista.com/statistics/273018/number-of-internet-users-worldwide/.

The History of Computing Project. "History of Computing Industrial Era (1985–1990)." March 20, 2006.

The New York Times. "Napster Documentary: Culture of Free | Retro Report", 2014.

Turner, Benjamin; Nutt, Christian *"Building the Ultimate Game Machine."* Nintendo Famicom, 2003

Chapter 13

Anderson, Chris. "How web video powers global innovation." *TED*

Bloomberg Business Week. "A New World Economy" August 22, 2005

Elgin, Ben. "Google Buys Android for Its Mobile Arsenal." *Bloomberg Businessweek,* 2005

Gary Rivlin. "Wallflower at the Web Party." October 15, 2006. New York Times.

Gillette, Felix. "The Rise and Inglorious Fall of Myspace." *Bloomberg Businessweek,* 2011

Haarstad H, Fløysand A "Globalization and the power of rescaled narratives: A case of opposition to mining in Tambogrande, Peru." *Political Geography,* 2007

Jeffries, Adrianne. "Disconnect: why Andy Rubin and Android called it quits." *The Verge,* 2013

Lapinski, Trent. "MySpace: The Business of Spam 2.0 (Exhaustive Edition)." *ValleyWag.* 2006

Merchant, Brian. *"The One Device: The Secret History of the iPhone."* 2017

The Daily Telegraph. "YouTube: a history"

Cnet.com. "Blockbuster laughed at Netflix partnership offer." http://www.cnet.com/news/blockbuster-laughed-at-netflix-partnership-offer/.

Chapter 14

Aaronson, Scott. "Why Electronic Music Rules." *Why Electronic Music Rules BBC,* 2010

ABC. "Printing breakthrough enables human sized functioning tissues." 2016. http://www.abc.net.au/news/2016-02-18/3d-printing-breakthrough-enables-human-sized-functioning-tissues/7178800

Bloomberg.com. "Why quantum computers will be super awesome someday." 2018. https://www.bloomberg.com/news/articles/2018-06-29/why-quantum-computers-will-be-super-awesome-someday-quicktake

Fortune.com. "Crispr made cheaper." 2018. http://fortune.com/2018/07/12/cancer-gene-editing-crispr/.

Geordie, Rose. "Quantum Computing: Artificial Intelligence Is Here." *www.cosomosmagazine.com*

https://cosmosmagazine.com/physics/quantum-computing-for-the-qubit-curious

https://medium.com/quantum-bits/what-s-the-difference-between-quantum-annealing-and-universal-gate-quantum-computers-c5e5099175a1

https://www.forbes.com/sites/bernardmarr/2017/07/10/6-practical-examples-of-how-quantum-computing-will-change-our-world/#65d3afd980c

Ladizinsky, Eric. Quantum AI Lab Google, LA *"Evolving Scalable Quantum Computers."* March 5, 2014.

Marr, Bernard. "Practical examples of how quantum computing will change our world." *www.forbes.com*

McDougall, Paul. "iPad Is Top Selling Tech Gadget Ever." www.census.gov, 1984 www.census.gov/hhes/computer/files/1984/p23-155.pdf

Medium.com. "Nurture AI." 2018. https://medium.com/nurture-ai/keeping-up-with-the-gans-66e89343b46

Medium.com. "What's the difference between quantum annealing and universal gate quantum computers."

New York Times. "Electric cars batteries." 2017. https://www.nytimes.com/2017/07/08/climate/electric-cars-batteries.html

Oxford University Press, USA. *"Superintelligence: Paths, Dangers, Strategies."* 2014

Phys.org. "Silicon gate quantum." 2017. https://phys.org/news/2017-12-silicon-gate-quantum.html

Pwc.com. "Impact of automation on jobs." https://www.pwc.com/hu/hu/kiadvanyok/assets/pdf/impact_of_automation_on_jobs.pdf

SITN. "Crispr, a game changing genetic engineering technique." 2014. http://sitn.hms.harvard.edu/flash/2014/crispr-a-game-changing-genetic-engineering-technique/.

Spacenews.com. "NASA selects Boeing and Space-X for commercial crew contracts." https://spacenews.com/41891nasa-selects-boeing-and-spacex-for-commercial-crew-contracts/.

Spectrum. "Intels new path to quantum computing." https://spectrum.ieee.org/nanoclast/computing/hardware/intels-new-path-to-quantum-computing#qaTopicTwo

Techcrunch.com. "Space-X aims to make history 3 more times in 2018." 2018. https://techcrunch.com/2018/02/09/spacex-aims-to-make-history-3-more-times-in-2018/.

Time. "Space-X ten things to know." http://time.com/space-x-ten-things-to-know

Towardsdatascience.com. "The most complete chart of neural networks explained." 2018. https://towardsdatascience.com/the-mostly-complete-chart-of-neural-networks-explained-3fb6f2367464

Usblogs.pwc.com. "Top 10 AI tech trends for 2018." 2018. http://usblogs.pwc.com/emerging-technology/top-10-ai-tech-trends-for-2018

About the Author

Dagogo Altraide is the founder of *ColdFusion*, the popular YouTube channel with over one million subscribers. *ColdFusion* creates cutting-edge documentaries and infotainment with themes based on the past, present, and future of tech-science. Altraide's work has been featured on Time.com and has been shown in classrooms and lecture halls worldwide. After graduating from Mechanical Engineering at the University of Western Australia with a graduate degree from Curtin University, Dagogo's mission has been to reveal the bigger picture of our world by making complicated topics easy to understand, often through a narrative lens. Interested in the stretching of minds, *ColdFusion* aims to connect, educate, and inspire the thinkers within us.

CPSIA information can be obtained
at www.ICGtesting.com
Printed in the USA
JSHW031943180521
14791JS00002BA/5

9 781642 505917